TOY SOLDIERS

IDENTIFICATION
AND
PRICE GUIDE

Other **CONFIDENT COLLECTOR** *Titles*
of Interest from Avon Books

ART DECO
IDENTIFICATION AND PRICE GUIDE
by Tony Fusco

BOTTLES
IDENTIFICATION AND PRICE GUIDE
by Michael Polak

FIREHOUSE MEMORABILIA
IDENTIFICATION AND PRICE GUIDE
by James Piatti

POSTERS
IDENTIFICATION AND PRICE GUIDE
by Tony Fusco

AMERICAN POTTERY AND PORCELAIN
IDENTIFICATION AND PRICE GUIDE
by William C. Ketchum, Jr.

SALT AND PEPPER SHAKERS
IDENTIFICATION AND PRICE GUIDE
by Gideon Bosker and Lena Lencek

WESTERN MEMORABILIA
IDENTIFICATION AND PRICE GUIDE
by William C. Ketchum, Jr.

40S AND 50S
DESIGNS AND MEMORABILIA
IDENTIFICATION AND PRICE GUIDE
by Anne Gilbert

60S AND 70S
DESIGNS AND MEMORABILIA
IDENTIFICATION AND PRICE GUIDE
by Anne Gilbert

TOY SOLDIERS

IDENTIFICATION AND PRICE GUIDE

FIRST EDITION

BERTEL BRUUN

The CONFIDENT COLLECTOR™

AVON BOOKS ◆ NEW YORK

Important Notice: All of the information, including valuations, in this book has been compiled from the most reliable sources, and every effort has been made to eliminate errors and questionable data. Nevertheless, the possibility of error always exists in a work of such scope. The publisher and the author will not be held responsible for losses which may occur in the purchase, sale, or other transaction of property because of information contained herein. Readers who feel they have discovered errors are invited to *write* the author in care of Avon Books so that the errors may be corrected in subsequent editions.

THE CONFIDENT COLLECTOR: TOY SOLDIERS IDENTIFICATION AND PRICE GUIDE (1st edition) is an original publication of Avon Books. This edition has never before appeared in book form.

AVON BOOKS
A division of
The Hearst Corporation
1350 Avenue of the Americas
New York, New York 10019

Copyright © 1994 by Bertel Bruun
Front cover photograph by Brian Edwards
Inside back cover author photograph by Ruth Bruun
Interior text design by Suzanne H. Holt
The Confident Collector and its logo are trademarked properties of Avon Books.
Published by arrangement with the author
Library of Congress Catalog Card Number: 94-1646
ISBN: 0-380-77128-4

Library of Congress Cataloging in Publication Data:

Bruun, Bertel.
  The confident collector : toy soldiers identification and price guide / Bertel Bruun.
    p.   cm.
Includes bibliographical references
1. Military miniatures—Collectors and collecting—Catalogs.
I. Title.
NK8475.M5B78   1994   94-1646
668.7'2—dc20   CIP

First Avon Books Trade Printing: August 1994

AVON TRADEMARK REG. U.S. PAT. OFF. AND IN OTHER COUNTRIES, MARCA REGISTRADA, HECHO EN U.S.A.

Printed in the U.S.A.

OPM 10 9 8 7 6 5 4 3 2 1

*This book is dedicated to my wife
and to all other patient and understanding spouses
of toy soldier collectors*

Contents

Foreword

✤ Some twenty-five years ago as I was going through an old box from my childhood, I came across a battered Lineol soldier that brought back a flood of childhood memories. A closer inspection revealed the craftsmanship with which he was made. I was hooked! Ever since that day I have collected toy soldiers and have tried to learn as much about the hobby as I could. Most of this knowledge has come from others. Although some collectors regard their expertise as a treasure to be jealously guarded, most are willing—and even eager— to share their knowledge. It is difficult to list all those who have helped me obtain the information contained in the pages of this book. Among the many who have been exceptionally helpful are Steve Balkin of Burlington Antique Toys, who has supplied me with many soldiers and with an even greater wealth of information; Peter Blum of the Soldier Shop, who has advised me wisely over the many years I have frequented his store; Ernie Capek, who suffered through some of this book's chapters as ideas were bounced off his level head and part of whose collection features in the photographs; Ernie's wife Mary, who let me take photographs of figures from her collection; Tony Grecco, who has given freely of his fund of knowledge and helped catch the most glaring errors of this book before the manuscript even reached the editor and who has let me photograph some of his soldiers; Henry Kurtz, formerly of Phillips, New York, now proprietor of his own auction house, who has spent many an hour teaching me about toy soldiers and who has supplied me with some of the photographs; Tom Scott, who let me photograph some of his figures and who gave freely of his knowledge of Lucotte and Mignot; Ernst Schnug of Germany and Egon Finsted of Denmark, who have patiently answered a multitude of questions about composition figures; and Lasse Peitersen, Jørgen Victor Hansen, and Helge Sheunchen—all Danes— who have helped unearth obscure information and figures. I also want to thank Katherine Montague, who brought my illegible notes into a semblance of readable information and my wife, Ruth, who not only has let me indulge in my hobby but has encouraged it and assisted expertly in making my English comprehensible. With all this help I am sure errors will still be found in this book and for these I take sole and full responsibility.

Bertel Bruun
Westhampton, September 1993

Introduction

❧ Miniature soldiers have been around for a long, long time. Some of the pharaohs of Egypt were entombed with rows of small soldiers carved from wood to protect and guard them in the next world. Such soldiers are on display in the Cairo Museum. An ultimate example of such an army can be found in Xian, China, where thousands upon thousands of almost life-size soldiers and horses made of terra-cotta have been—and are still—being unearthed. However, these armies were not considered toys, although similar figures might indeed have been used as toys.

It is known that male children of royalty and noblemen in times past had toy soldiers made for them in metals, precious as well as base. These soldiers served to illustrate and, to some degree, teach boys the intricacies of tactics in what was to be their main pursuit in life—war.

The first mass-produced toy soldiers—the subjects of this book—appeared in the early eighteenth century in Germany. These were the flats made by silver and pewtersmiths in Germany. Although pewter was first used, it was soon replaced by lead. These soldiers, greatly detailed and inexpensive, became immensely popular in continental Europe and millions of them were produced. Production on a smaller scale has continued to this day. At about the same time, soldiers made of paper were introduced. These remained the most inexpensive of all and were especially popular after chromolithography was invented in the nineteenth century. A small production in different parts of the world still continues.

Fully round, mass-produced lead soldiers were first made in France at the time of the French Revolution. The first known maker was Lucotte, whose molds have survived in the possession

Models of the almost life-size soldier and his horse, from the emperor's tomb in Xian, China. Such models are sold as souvenirs.

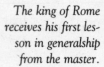

The king of Rome receives his first lesson in generalship from the master.

of CBG Mignot. In Germany flats were filled out to some degree (semi-flats), a type of soldier that was popular because it lent itself to home castings. Some companies specialized in selling mold kits. Not until fear of lead poisoning became a concern in the first half of the twentieth century did home casting lose its popularity.

The fully round figures were the most pleasing and in the middle of the nineteenth century, several German companies took up the production. Foremost among them Gustav Heyde, who, with an obvious flair for business, swamped the Western world with his millions of little soldiers.

In the late 1800s William Britain introduced hollow casting to toy soldier manufacturing. The advantage of the process, which reduced the amount of lead in each figure thus lowering the cost of both raw material and transport, was obvious and the process was taken up by many others, especially British companies. At the same time, on continental Europe, composition, a mixture of sawdust, glue, and kaolin, found its place it held until the advent of plastic.

The children of the United States largely depended on imported toy soldiers until the 1920s and 1930s, when the so-called dime-store figures were introduced. These figures are uniquely American, finding only a few imitators abroad. They took over a large part of the market in the United States and are consequently the most popularly collected type here.

And then came plastic, cheap, nontoxic, and light; the ideal material for toy soldiers. Between 1940 and 1960 plastic almost completely replaced all other materials in toy-soldier production.

With the demise of the old lead soldiers and an increasing interest in them, a new business sprung up in the 1970s, the production of so-called new old toy soldiers. These figures are not meant as toys but as collector's items, either by themselves or in combination with original toy soldiers.

The popularity of toy soldier collecting is relatively new. A few prescient persons started their collections long before production of lead soldiers ceased, but these lucky ones are indeed few in number.

The first toy soldier auction was held at Knight, Frank and Rutley in England in 1968. Since then such auctions have become regular events throughout the Western world.

Special shows are being held in increasing numbers and places both in the United States and Europe. Periodicals printed, books published, and newspaper articles covering toy soldier events are all proof of the growth of the hobby.

Toy soldier collectors are mainly, but far from exclusively, men. Nostalgia plays an important part in this. Handling the toys of one's youth is indeed a pleasurable experience. Other aspects such as historical and/or military interest also play a role, as does the "pack rat instinct" characteristic of almost all collectors regardless of subject.

Most toy soldier collectors limit their collections in one way or the other. They may collect soldiers produced by a single firm (i.e., Britains), in a special period (i.e., pre World War I), of a special service (i.e., Air Force), groups of a special configuration (i.e., bands), made of a special material (i.e., paper), and so on.

Pre-war Britains and pre- and postwar Elastolin catalogs. Such catalogs are of great interest to the collector of toy soldiers.

At the present time in the United States the most popular objects of toy soldier collectors are dime-store figures and Britains.

Knowledge about toy soldiers is attained in different ways. Several books, both general as the present or more specialized, are available (see Bibliography, page 349). Catalogs put out by the producer, either original or in reprints, are rich sources of information. Such catalogs are often available from dealers or photocopies might be obtained. Dealers are often willing to answer questions as are other collectors. A subscription to one or more of the periodicals listed in Appendix A, page 337, keeps one abreast of new discoveries as well as of stores and auctions. Most knowledge though comes from seeing and/or handling as large a variety of soldiers within one's sphere of interest as possible.

Collectors find their objects on interest in several different ways. Antique shops and shows, flea markets, and garage sales sometimes can lead to major finds, but as the hobby is getting better known, chances of the "deal of your life" have greatly diminished. Specialty stores and mail order vendors (see page 341) offer a far better choice. Specialty auctions (see page 339) are another venue for enlarging collections. The prices at such auctions are, however, very variable. Remember it takes only two bidders to send the price of a given object through the ceiling and once both of these theoretical bidders have acquired the soldier of their desire it may have no other takers. Unfortunately, when prices inflated in this manner are reported in the press, many people erroneously assume that similar soldiers in their possession are of great value. Nevertheless, specialty auctions also sometimes offer real bargains and it is always worthwhile seeing the many toy soldiers collected in one place for the event.

The condition of a toy soldier or set is of the greatest importance in establishing its value. Although varying somewhat in detail, the following scale for describing condition is used:

Mint   Perfect condition in every way and never taken out of its box

Excellent   No obvious sign of any damage

Very good   Some evidence of being played with, with minor scratches and/or imperfections

Good   This soldier has seen action but is still intact with most of the paint extant

Fair   This is a veteran of many wars who has lost considerable paint and may have minor imperfections

Poor   Anything below Fair

Pluses and minuses may be added (except of course to "Mint") to further fine tune the condition.

This type of description is particularly important when buying from mailing lists or absentee auctions.

Especially for Britains the presence of the original box is very important, enhancing the value by 10 to 25 percent or even more. The condition of the box is also important in establishing a set's price.

As in many other fields of collection, rarity is a very relative term. A rare set of Britains for which there is a large demand is indeed valuable. However, a set of a little known or little collected manufacturer may be far more rare in absolute terms while its actual value is but a fraction of that of the "rare" Britains set.

How to Use this Book   The present book is an attempt to help the beginning collector as well as dealers who encounter toy soldiers identify and make a reasonable evaluation of the soldiers they have encountered. Identification is the first step. The book is divided into sections according to the materials with which the soldiers are made: lead, composition, iron, etc. The section on lead, the biggest, is further subdivided into flat, semi-flat, solid, and hollow-cast groupings. In each section different manufacturers and possible special markings of their productions are described and examples of prices given. These prices are based on the condition being good or very good, but not mint.

I have tried to give examples of prices for a range of rarities of products, but no attempt has been made to cover all toy soldiers produced. Nevertheless it should be possible to reach a reasonable evaluation of a given soldier or set of soldiers in hand. The photographs have been chosen to illustrate as large a range of products as possible but are by no means all inclusive.

Dimensions are given in both inches and millimeters or centimeters. The unit of measurement which is standard in the country of manufacture or among collectors is given first.

In the back are sections describing other books which might be helpful, periodicals, auction houses, and dealers specializing in old toy soldiers.

A special section covers "new old toy soldiers," an increasingly larger part of the hobby, although not strictly *toy* soldiers. Likewise sections on vehicles, cannons, and airplanes, as well as fortifications and castles used in conjunction with toy soldiers are included.

Enjoy!

I.

Lead

❧ Until the advent of plastic, lead was by far the most commonly used material in the making of toy soldiers. Alloyed with relatively small amounts of other metals, the hardness of the product can be controlled and the relatively low melting point makes handling easy. The softness of the metal makes it possible to bend and cut the figure at will after it is cast and also to solder, making it easy to attach equipment of various sorts. The drawbacks are the price and availability of lead. During times of war the former increases while the latter plummets. The figures are subject to lead disease (or lead rot), a process of oxidation turning the lead into white dust. The cause of this problem seems to be related to casting temperatures as well as pretreatment of the figure before painting (they should be varnished before a basecoat is applied). The high weight makes shipping costs high. Although there is increasing awareness of the danger of lead toxicity, it is worth noting that there are no reports of lead poisoning in either children or adult collectors who handle lead soldiers.

For the sake of convenience lead soldiers can be divided into flats (see below), semi-flats (see page 16), solids (see page 19, and hollow-cast figures (see page 90). Although there might be an overlap, most companies specialized in making one or the other of these types of lead figures and collectors tend also to concentrate their efforts on one of these groups.

In the 1950s plastic took over as the leading substance for toy soldier making and today only new old toy soldiers (see page 196) and reissues of flats are made in this material. Pending legislation may stop even this very limited production, but it seems possible that for most purposes high lead content alloys can be replaced by alloys containing far less or no lead at all.

FLATS

So-called flats were the first type of toy soldiers produced *en masse*. By definition, flats are two-dimensional metal figures less than 1 millimeter (0.04 inch) in thickness. The molds were made in slate, the two sides having the front and back, respectively, engraved on them. A small hole allowed the melted metal to be poured into the mold. Originally the metal used was pewter (*zinn* in German, thus the term *zinnfiguren* is often used for these soldiers), but soon an increasing amount of lead was added to make the pouring easier and less costly.

Traditionally the father of flats is considered to be the German pewtersmith Johann Gottfried Hilpert who set up shop in Nürnberg in the latter part of the eighteenth century. Other tinsmiths in Germany and other countries soon followed with their own production and a plethora of early manufacturers are known. The early figures tended to be large, 5 to 10 centimeters

(2 to 4 inches) tall. Some were even given movable arms. The subjects depicted were contemporary military, as well as historical (especially knights), civilian, fairy tale and religious figures, and animals. Because of their thinness, very little metal was used, which when combined with the low wages for painters allowed the figures to be sold very inexpensively. The figures were sold in sets packed in wooden boxes, either a square or a thin-walled oval variety. They were packed with excelsior for protection. For the first fifty years of this production, no standard size was used and figures varied widely from one company to the next as well as within a given company's production. This latter phenomenon was at least partially caused by the trade in molds. The companies were small and often lasted only a short while. As the molds were made of stone, they were virtually indestructible and their only possible value lay in being sold to other masters.

As the business matured the market for the products covered all of Europe as well as the United States. The center for production however remained in Germany and fewer companies dominated the market. The largest of these was Ernst Heinrichsen, which dominated the production for almost a century (it stopped production during World War II). Heinrichsen standardized the size to 28 millimeters (1⅛ inches) for standing foot figures, a size soon accepted by other makers as well. Flats remained the most popular toy soldiers until the latter part of the nineteenth century when full-bodied figures began to take over. Although the dominance of the market was lost, production of flats continued until World War II (Ochel of Kiel, Germany, that produced their Kilia range until well after the close of the war). In the twentieth century the packing changed to cardboard boxes, with paper padding.

There has been a certain overlap of producers of flats and solid lead figures. Thus Mignot of France, best known for its beautiful solids, produced a number of beautiful flats. Production on a small scale—now meant for collectors and not children—continues to this day, both in Germany and Austria.

Early, large flat of
mounted knights;
value $30.

Large flat of postal
wagon; value $25.

Early flat of various large
sizes; value $12 each.

Beautifully detailed large
flat; value $25.

Identification   If not in their original boxes, flats are very difficult to identify. A few are marked on the base but the vast majority are unmarked. Some companies tended to use certain foot shapes for figures but this is far from reliable as an identification feature (the trade of molds among makers does not help either). The term "Heinrichsen-style figures" has become common because of this difficulty in identification. Assignment of individual pieces to specific makers is best left to the specialists.

Prices   The market in the United States for flats is quite small. As collectors' subjects they are more popular in continental Europe, Germany in particular. The prices tend to be low, varying from less than a dollar to $5 per piece. Larger and more complex flats may, of course, be more expensive.

Early wooden box for flats.

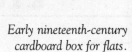

Early nineteenth-century cardboard box for flats.

Mid-twentieth–century cardboard box for flats.

Top row: Napoleonic
camel corps by Mignot;
value $75. Rows 2 to
5: Mounted Arabs,
probably by Heinrichsen;
value $75. Right: British
camel corps at rest,
probably by Heinrichsen;
value $25.

Flats of various
armies by
Heinrichsen; value
$25 per row.

Heinrichsen flats.
Top: Riding school;
value $50. Middle:
Mounted staff officers;
value $25. Bottom:
Prussians attacking;
value $30.

Lancer
encampment by
Heinrichsen;
value $50.

Medieval warriors
by unknown
maker; value $75.

German, French,
and Austrian Great
War soldiers by
Kilia; value $25
per set (2 rows).

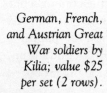

Boxed sets (usually about twenty-five foot figures or fifteen mounted figures) sell for $20 to $50 per set, depending on age, quality, and subject. Most desirable are Heinrichsen boxes. Others such as Kilia figures find few customers and, therefore, may be even cheaper.

SEMI-FLATS

An attempt to improve on the flats was made in the latter part of the nineteenth century. This resulted in a taller and thicker figure than the ordinary flat, a so-called semi-round or semi-flat toy soldier. Such figures, really neither fish nor fowl, reached some degree of popularity and production has continued on a small scale. Generally, semi-flats lack the detail of the flat and do not attain the three-dimensional reality of the solids. One advantage, though, is that semi-flats are easily cast. This advantage was capitalized on by companies that sold molds (made in metal) for home casting. Foremost among these mold producers was the German company Gebrüder Schneider, whose molds can be found throughout Europe and the United States. Home casting went out of fashion in the 1930s when the possibility of lead poisoning was recognized. Several companies, such as Schweizer and Kaler, whose main production was flats, experimented with semi-flats. Solid soldier manufacturers such as Mignot and Heyde have also experimented with semi-flats. Needless to say such factory-made figures are of a much higher quality than the home-cast varieties.

Identification Semi-flats are more than 1 millimeter (0.04 inch) thick, but not fully three dimensional. As with flats, the manufacturer's name is occasionally engraved on the foot, but most are unmarked. Attributing isolated pieces to individual manufacturers is extremely difficult except for true connoisseurs. Home-cast figures are usually quite easily recognized both by the coarseness of casting as well as the crudeness of painting. Although adequately fulfilling their function in a child's army, they lack the charm and quality of their professionally produced cousins.

Factory-painted semiflats: British Line Infantry; value $2 each.

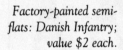

Factory-painted semiflats: Danish Infantry; value $2 each.

Factory-painted semiflats of high quality: British Life Guard; value $25.

Unpainted semiflats, probably factory cast; value $1 and $2.

Homecast semiflats, crude but functional; value $2 each.

Homecast and painted semiflats from Schneider molds; value $2 each.

Homecast and painted semiflats; value $3 each.

Prices Semi-flats have an even smaller following than flats and most pieces, unless exceptional in either complexity and/or painting, tend to sell in the $1 to $3 range. Home-cast figures should not cost more than a dollar.

SOLIDS

Fully round, three-dimensional toy soldiers cast in solid metal, usually a lead alloy, are known as solids. These, together with their derivatives, the hollow casts, are the most popular soldiers among collectors.

The glory and the honor of introducing solid figures falls to the French. It was in revolutionary France that the mass production of these soldiers started and it was here that the basic scale of metal soldiers, 54 millimeters (2⅛ inches), was established. Although several companies existed in Paris at the time, the most successful of these was Lucotte, later absorbed by the company of Mignot. These companies have concentrated their production on French historical figures as well as the armies of the First Empire and later French soldiers. Only relatively few sets of foreign troops (including U.S. Civil War figures and doughboys) have been produced in France.

Although much more expensive than flats, solids were successful enough to tempt German makers into the field in the middle of the nineteenth century. Here, however, production was not just for the home market, but a strong export market was established leading to a much greater variation in the subjects chosen. The German companies also experimented with different sizes (see below). The alloy used contained a higher concentration of lead than the French alloy, making it possible to bend the limbs into virtually any position desired. This flexibility made it possible to depict many more and varied activities, clearly distinguishing these German soldiers from the far more staid Mignot figures and the later Britains hollow casts. Solids were also made in other countries, Italy, Spain, and Denmark among them. Even in the United States a number of companies producing solids sprang up at the turn of the century. None of these companies though ever approached the larger German companies, especially Heyde, in output.

In the last twenty-five years solids have again been produced, not as toys but as collectors' items, the so-called new old toy soldiers (see page 196).

✤ LUCOTTE/MIGNOT

The oldest maker of solids was Lucotte. Around the year 1789 in Paris as noble and ignoble heads were falling, Lucotte (known by that one name only) started producing what may be not only the first but possibly also the best solid-cast lead soldiers. Napoleonic regiments and armaments, cannon, field forges, as well as the emperor himself with his generals formed the main output although other types, for instance World War I poilus, were also made.

Lucotte became absorbed into Mignot in 1928 but the latter company continued to issue at least some of the Lucotte sets after having taken over the molds. Mignot (originally Cuperly, Blondel & Gerbeau or CBG) was probably formed in 1825 to produce solid lead soldiers, quite similar to Lucotte's (see Identification). The name changed to Gerbeau and Mignot in 1900 when Henri Mignot joined the company of which the Gerbeau family now were the sole owners. When the name Gerbeau was dropped and the name Mignot stood alone is not clear, but the figures produced by any of these companies are known as Mignot figures.

From the beginning both Lucotte and Mignot soldiers were expensive and sales seemed to have been mainly limited to France. Only late in the second quarter of the twentieth century did Mignot start catering to foreign demands in a half-hearted fashion. At that time, a number of U.S. Civil War, West Point cadets, and doughboy sets were produced, but export never came to be of major importance for Mignot as it was for Heyde in Germany and Britains in England.

The production of Mignots seems to have been a somewhat haphazard affair and certainly appears never to have left the preindustrial stage in which it originated. After much struggle the company has now closed its doors and gone into receivership.

Identification Both Lucotte and Mignot are among the aristocrats of toy soldiers. They are extremely well made, in regard both to casting and painting.

Lucottes, which are much rarer than Mignots, will occasionally have the foot stamped with the Imperial Bee (its trademark) and/or the letters LC or LP, the later probably standing for Lucotte, Paris. Some Mignots have paper labels marked MADE IN FRANCE, others (only postwar) are marked CBG, MADE IN FRANCE or MADE IN FRANCE on their square bases, but most are unmarked. The styles of both companies are, however, characteristic: the heads are plugged in; the arms are fixed, although in many cases, they attached to the body after casting; guns and musical instruments are usually soldered on and not part of the original casting. Mounted figures are detachable. The mounted figure does not have the characteristic spike for attachment that Heyde and most other solid figure makers used. Naturally the horses lack holes for such spikes.

Distinguishing between Lucotte and the far more common Mignots is important. Lucotte's foot figures march forward with a longer stride, more energetically than Mignot's, which seem nearer the end than the beginning of their march. The base is therefore longer for Lucotte figures than for Mignots. Contrary to most other soldier makers, many marching Mignot figures have the right rather than the left foot forward. Mounted figures are best distinguished by the horse. Lucotte horses are more elegant than Mignot's and have longer tails (Mignot's walking horse has the tail cropped short). Whereas Mignot horses have saddles and/or saddle blankets as an integral part of the casting of the horse, Lucotte's saddle blankets and saddles with attached stirrups (which are usually too long) are detachable.

Mignot infantry was issued in red boxes with twelve (or later eight) figures usually including a standard bearer, officer, and bugler. Cavalry sets contained five mounted men. For export to the United States, six mounted men were included in a box for a certain period. In the 1960s, Mignot issued boxed sets of only four foot figures: officer, standard bearer, bugler, and private. Personality figures were issued either singly, in pairs (Louis XVI and Marie Antoinette), or in groups (Napoleon and his staff).

Sometimes these figures were mounted on a small wooden base with the name of the personality.

Characteristically, Mignot (and rarely Lucotte) issued diorama boxes. These may contain military or civilian subjects and are often very beautiful, richly adorned with furnishings such as bushes and trees (several hunt scenes) or furniture and equipment (cavalry garrison with stable and sleeping quarters for the men). The labels in Mignot boxes varied somewhat with time, but the content is usually written in hand or rubber-stamped in the center.

It is difficult to determine the age of a given Mignot figure, but paint styles and base colors are helpful.

Until about 1930, the base color was dark green. From the early thirties to about 1970 the base was a light sand color and after 1970 a very dark brown color. The paint work is generally better executed on earlier figures; this is especially noticeable in the facial features. In the 1950s, Mignot introduced a full-gloss paint (older figures are semi-gloss) but in the 1960s switched to a matte type of paint, only to go back to semi-gloss in the very late 1980s.

Prices Lucotte and Mignot were expensive toys — and they still are! Because of this they have tended to be treated more respectfully by their young owners and relatively many have survived as boxed sets. When encountered singly, Lucottes

Lucotte: mounted French officer; value $100.

sell for about $30 to $50 for foot figures, $75 to $100 for mounted figures. Mignots are less expensive, $15 to $25 for foot figures and $25 to $35 for mounted figures.

Most Mignots are encountered boxed and in excellent to mint condition. The list below is by no means complete but illustrates the approximate prices Lucottes (usually unboxed) and Mignots (boxed) have attained at auction in the last few years. On the average, the regular six-piece cavalry set and twelve-piece infantry set sell for around $225 in mint and boxed condition.

Lucotte: mounted British officer. Note horse's long tail and separate saddle, saddle blanket, and stirrups. Value $100.

Lucotte: marching French soldiers. Note long stride. Value $50.

Lucotte: French soldiers in action; value $50 each.

Lucotte: doughboy kneeling firing; value $50.

Lucotte British soldiers in action; value $50 each.

Mignot: French musketeers of Louis XIII; value $200. Photograph courtesy of Henry Kurtz Ltd.

Mignot: Israeli Infantry on parade; value $300. Photograph courtesy of Henry Kurtz Ltd.

Mignot: Infantry of Henry IV; value $200.

Mignot: Napoleonic ammunition wagon; value $250.

Mignot: Napoleon and his staff; value $500.

Mignot: French poilu in action; value $25 each.

Mignot: Chinese
and early eighteenth-
century soldiers.

Mignot:
Monaco
troops.

Mignot: French
artillerists.

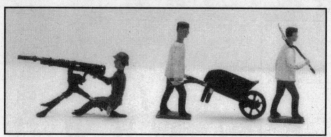

Mignot: French machinegunner and work crew.

Mignot:
Thai soldiers.

Mignot: Israelis in action.

Mignot:
Union
artillerists.

Mignot:
Confederates
marching
and Union
Infantry.

Mignot: *examples of cavalry.*

Mignot: *examples of cavalry.*

Mignot: *examples of cavalry.*

Mignot: *examples of cavalry.*

*Mignot: examples
of cavalry.*

*Mignot: example
of cavalerist on
rearing horse.*

Mignot: examples of ancient and near-ancient soldiers.

*Mignot:
examples
of ancient
soldiers.*

Mignot:
examples of
ancient and
near-ancient
soldiers.

Mignot:
examples of
Napoleonic
soldiers.

Mignot:
examples of
Napoleonic
soldiers.

Mignot:
examples of
Napoleonic
soldiers.

Mignot: *a medley of figures.* Note boy scout on right.

Mignot: *a medley of soldiers.*

Mignot: French soldiers.

Mignot: French soldiers.

Mignot: St. Louis
and Louis XI.

Mignot:
Richelieu.

Mignot:
Charles II.

Mignot: Cleopatra,
Pius VII, and Marie
Antoinette.

Mignot: King George
V and Queen Mary.

Mignot: Richard
the Lionhearted,
the Dauphin,
Charles I, and
Lafayette.

Mignot: Marshall
Joffre, Napoleon, and
Joan of Arc.

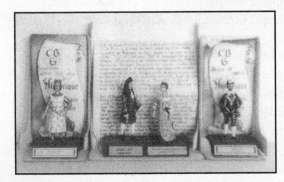

Mignot: personali-
ties on wooden
stands: St. Louis,
Louis XIV and
Madame de
Maintenon, and
Henri III.

Mignot: Napoleon and his staff.

Lucotte

Mounted

Grenadier à cheval (7 pieces)	$960
5th French Hussars (10 pieces)	$900
Vistula Lancers (6 pieces)	$600
French Carabiniers (6 pieces)	$500
Napoleon and his General Staff (6 pieces)	$825
The Empress Josephine's Dragoons, 1812 (9 pieces)	$2,000
French foot Napoleonic Hussars (9 pieces)	$500

Foot

3rd Swiss Regiment Fusiliers marching (12 pieces)	$660
3rd Swiss Regiment of the Guard marching (7 pieces)	$880
Fusiliers of the Line marching (12 pieces)	$500
Grenadiers of the Vistula Legion (9 pieces)	$500
Dutch Grenadiers of the Guard (12 pieces)	$600
Flankers of the Imperial Guard (24 pieces)	$1,200
Valtigeurs of the Guard (16 pieces)	$800
United States Infantry Doughboys firing (9 pieces)	$400

Imperial Guard Grenadiers marching
   (12 pieces) $1,800
Band of the 3rd Swiss Regiment (12 pieces) $1,100

Equipment
French Napoleonic Horsedrawn Artillery Caisson $700
French Napoleonic Field Artillery Gun Team $725
Spanish Army (1809), Horsedrawn Artillery $600

Mignot—Reissue of Lucotte

French Napoleonic Field Artillery Caisson $400
French Napoleonic Poste Chaise $400
Napoleon's Imperial Coronation Coach $2,000
Pope's Ceremonial Coach $1,200
Napoleon mounted $150
Marshall Bessiere mounted $150

Mignot

Cavalry sets—Napoleonic
Vistula Lancers, 1808 (6 pieces) $200
2nd Regiment, Light Cavalry, 1812 (6 pieces) $150
Elite Gendarmes, 1810 (6 pieces) $325
French Carabiniers, 1812 (5 pieces) $250
Prussian Hussars, 1813 (5 pieces) $200
Prussian "Death Head" Hussars, 1812 (5 pieces) $200
French Napoleonic Cuirassiers, 1809 (6 pieces) $200
French Napoleonic Camel Corps, 1799 (6 pieces) $250
French Napoleonic Guard of Honor, 1813
   (6 pieces) $225
Bavarian Uhlans, 1805 (6 pieces) $225
French 2nd Hussar Regiment, 1808 (6 pieces) $225
Polish Lancers, 1812 (6 pieces) $250
Austrian Cuirassiers, 1812 (6 pieces) $250
Dutch Lancers, 1812 (6 pieces) $225

Light Cavalry of the Guard, 1809 (6 pieces)	$250
Imperial Guard of Honor of Strasbourg, 1805 (6 pieces)	$225
Berg Lancers, 1809 (6 pieces)	$225
Austrian Hussars, 1813 (6 pieces)	$225
Spanish Cavalry, 1808 (6 pieces)	$200
Scouts of the Guard, 1813 (6 pieces)	$225
1st Life Guard, 1815 (5 pieces)	$200
5th French Hussar Regiment, 1805 (5 pieces)	$220
10th Regiment French Hussars, 1812 (6 pieces)	$230
2nd Regiment French Chevaux Legers, 1812 (6 pieces)	$200
Mamelukes, 1810 (6 pieces)	$275
Napoleonic Guides, 1805 (6 pieces)	$250
Imperial Guard of Honor of Strasbourg, 1805 (6 pieces)	$200
Bavarian Uhlans, 1805 (6 pieces)	$200
French Gendarmes d'Elite, 1810 (6 pieces)	$200
General Augereau's Guides, 1897 (6 pieces)	$225
French Field Artillery Guard, 1809 (6 pieces)	$225
French Chasseurs à Cheval, 1809 (6 pieces)	$225
11th Regiment French Hussars, 1812 (6 pieces)	$225
1st Regiment French Hussars, 1810 (5 pieces)	$250
2nd Regiment French Hussars, 1808 (5 pieces)	$330
4th Regiment Austrian Hussars, 1813 (6 pieces)	$130
7th Regiment French Hussars (6 pieces)	$375

Other Cavalry Sets—Non-Napoleonic

Russian Cossacks, 1900 (6 pieces)	$175
English Royal Horse Guard, 1900 (6 pieces)	$175
Prussian "Death Head" Hussars, 1910 (6 pieces)	$200
Prussian Uhlans, 1914 (6 pieces)	$550
French Army Cuirassiers, 1890–1910 (6 pieces)	$160
Fourteenth-Century Knights attacking (6 pieces)	$220
Ancient Roman Cavalry advancing (6 pieces)	$175

Russian Imperial Guard, 1890–1914 (6 pieces) $225
French Chasseurs à Cheval, 1890–1914 (6 pieces) $225
1st English Life Guards, 1815 (6 pieces) $200

Bands
Mounted Band of the Polish Lancers, 1810
   (11 pieces) $660

Foot—Napoleonic (12 pieces)
Imperial Guard of Honor of Strasbourg, 1805 $225
Isenburg Grenadiers, 1808 $200
Italian Light Infantry, Beauharnais Regiment,
   1810 $200
French Light Infantry Skirmishers, 1809 $200
Russian Infantry, 1812 $175
Engineers of the Guard attacking, 1812 $250
Foot Guards of Amsterdam, 1806 $200
Chasseurs of the Italian Guard, 1810 $1,250
Spanish Light Infantry, 1808 $190
Grenadiers of the Imperial Guard, 1812 $250
8th Bavarian Infantry Regiment, 1805 $160
Dutch Grenadiers of the Imperial Guard, 1812 $200
Imperial Guard of Honor of Lyon, 1805 $250
Russian Guard Grenadiers, 1812 $175
Vistula Legion, 1808 $220
5th Battalion Orphans of the Imperial Guard,
   1812 $220
French Dragoons of the Guard, 1812 $220
Grenadiers of the Imperial Guard attacking,
   1812 $330
Voltigeurs of the Guard, 1812—special paint $360
Flankers of the Guard, 1811–1814 $190
French Volunteer Infantry, 1790 $275
Voltigeurs of the 33rd Line Infantry Regiment,
   1806 $200
13th Prussian Regiment of the Line, 1807 $200

Foot Guard of Amsterdam, 1806	$200
Würtenberg Line Infantry, 1806	$175
Legion du Nord, 1806	$175
Battalion of Valais, 1806	$175
Prussian Infantry, 1805	$230
Infanterie Légère	$200
Regiment de la Tour d'Auvergne, 1806	$175
4th Swiss Regiment, 1810	$300
British First Regiment Grenadier Guards advancing, 1812	$200
British Infantry of the Line advancing	$275
French Chasseur de la Garde	$225
French Napoleonic Infantry of the Line, 1812	$330
Russian Infantry of the Line firing, 1812	$225
French Marines of the Imperial Guard, 1812	$200
Departmental Guard of Paris, 1810	$275

Other Foot—Non-Napoleonic (12 pieces)

French Foreign Legion, 1863	$200
Turkish Infantry attacking, 1910	$200
Italian Infantry, 1914	$175
Indian Army Infantry, 1910	$175
Prussian Infantry assaulting, 1914	$175
Indian Army Infantry attacking, 1910	$150
Infantry of Henry IV	$150
French Foreign Legion attacking, 1863	$320
French Foreign Legion shooting, 1863	$220
Thirteenth-century Archers in action	$200
Ancient Gauls attacking	$250
Ancient Franks	$250
Infantry of Louis XIV	$175
West Point Cadet in winter uniform	$220
Confederate Artillery Gunners	$150
Confederate Infantry work party	$300

Crusader Knights	$250
Saracen Infantry	$165
Ancient Romans advancing	$175
Ancient Egyptian Infantry	$175
Prussian Infantry, 1915	$175
New York and New Jersey Continental Army Regiments, 1779	$220
New England Regiment of the Continental Line, 1779	$200
British Grenadiers of the 33rd Regiment, 1778	$200
French Foreign Legion in action, 1906	$330
Arabs standing	$220
Infantry of Louis XIV	$175
Japanese Infantry, 1905	$250
Russian Infantry, 1905	$360
French Chasseurs à Pied, 1914	$250
French Marine Fusilliers, 1910	$200
French Infantry of the Line, 1910	$200
U.S. Infantry, 1917–1918	$200
Scottish Highlanders firing, 1900	$200
British Grenadiers, 1900	$175
Monaco Royal Guard	$600
Israeli Infantry in action	$365
Israeli Infantry in parade dress	$365
French Army work party, 1914	$180

Foot—Napoleonic (8 pieces)

Sappers and Drummers of the Grenadiers of the Imperial Guard, 1812	$150
Russian Infantry Grenadiers, 1807	$100
Austrian Infantry, 1800	$150
French Line Infantry, 1809	$165
Swiss Regiment, 1812	$150
French Napoleonic Dragoons marching	$130
Sappers of the Grenadiers of the Imperial Guard	$125

Other Foot—Non-Napoleonic (8 pieces)

French Army machine gunners, 1916 (9 pieces)	$310
Assyrians	$150

Foot—Napoleonic (4 pieces)

Dutch Grenadiers	$100
French Artillery servants	$100
Russian Grenadiers firing, 1807	$75
Austrian Infantry advancing, 1800	$75
Sappers and Grenadiers of the Imperial Guard	$75
Drummers and Grenadiers of the Imperial Guard	$75
Italian Grenadiers of the Imperial Guard	$75
Fusilliers of the Guard	$75
Voltigeurs of the Guard	$75
French Dragoons	$75
Marines of the Guard	$75
French Light Infantry	$75
French Artillery servants	$75
Imperial Guard bandsmen	$75

Other Foot—Non Napoleonic (4 pieces)

Infantry of Louis XIV	$40
Confederate work party	$40
French work party (1910)	$40

Bands (12 pieces)

Band of the Parisian Guard, 1910	$350
Band of the French Foreign Legion in khaki	$400
Band of the Imperial Guard Grenadier, 1812	$550
Band of the French Foreign Legion, 1906	$475
Band of the 17th Regiment of Chasseurs, 1809	$200
Band of the Legion du Nord, 1806	$250
Band of the Alpine Chasseurs, 1790–1910	$225
Band of the Dutch Grenadiers of the Imperial Guard, 1805	$250

Band of the U.S. Army, 1917	$250
Band of the Imperial Guard of Honor of Strasbourg, 1805	$300
Band of the Grenadiers of the Imperial Guard, 1812	$275
Band of the German Army, 1917	$250
Band of the 17th Regiment of Chasseurs, 1809	$300
Band of the French Infantry of the Line, 1910	$300
Band of the French Navy, 1910	$300
Band of the U.S. Infantry, 1917–1918	$300

Dioramas (with original box)

Piano recital at an evening party, 1900 (13 pieces)	$300
Two-tier French army telegraph display, 1900 (about 35 pieces)	$800
Bulgarian Infantry in action, 1910 (8 pieces)	$250
Hunt scene (about 30 pieces)	$310
Circus (8 pieces)	$300
French Cavalry Stable diorama	$250
Arab Caravan display (no box) (18 pieces)	$1,000
Three-tier North Africa display box (73 pieces)	$1,400
Napoleon's Farewell at Fontainebleau (16 pieces)	$475
"Roaring Twenties" street scene (11 pieces)	$275
"Roaring Twenties" Jazz band (6 pieces)	$650
Two-tier Joan of Arc display (40 pieces)	$1,500
Jupiter's chariot (6 pieces)	$300

Equipment—Military

Confederate horsedrawn gun team	$150
U.S. Army Pontoon section	$250
French Army Open Coach with four-horse team, 1900	$350

Equipment—Civilian
Open coach drawn by four-horse team, 1900 $300

Personalities—Mounted
Napoleon and his general staff (8 pieces) $500

Mignot made many other mounted personality figures. Below is a partial list.

French	Other
Charlemagne	Caesar
Joan of Arc	Attila
St. Louis	Richard the Lionhearted
Louis XI	Saladin
Louis XIII	General Robert E. Lee
Louis XIV	General Ulysses S. Grant
Louis XV	Charles I
Henri III	
Henri IV	
François I	
Napoleon I	
Napoleon III	
Marshall Foch	
Marshall Joffre	
Marshal Lyantey	

These mounted figures cost about $50 in mint condition.

Personalities—Foot
Napoleon in Chasseur uniform $75
Napoleon in greatcoat $75

Mignot made many other personality figures. The following is a partial list.

French	Other
Jeanne d'Arc	George Washington
St. Louis	Abraham Lincoln
Louis XI	Charles I of England
Louis XIV	Queen Victoria
Louis XVI	King George of England
Louis XVII	Queen Mary of England
Henri III	Richard the Lionhearted
Henri IV	Saladin
Richelieu	Ramses II
Lafayette	Cleopatra
Napoleon	Christopher Columbus
Marshal Joffre	Mary Stuart
Madame de Maintenon	Pope Pius VII
Madame de Montespan	
Marie Antoinette	
Empress Josephine	

These personality figures cost about $35 each in mint condition.

♣ HEYDE

When Gustav Adolf Theodor Heyde established his company in the middle of the nineteenth century his was but one among many firms producing solid lead soldiers in Germany. Others, such as Conrad Scheild Knecht, Georg Spenküch, and Johann Haffner had also started production of solids (some working their way from flats through semi-flats to solids). By about 1870, however, when Georg Heyde had taken over the leadership from his father, the company was well ahead of all its competitors. It remained so until its violent demise in the inferno of saturation bombs released over Dresden by RAF bombers between February 13 and 15, 1945. In that raid the factory and its records disappeared. But the 1870s were happier days when the United Germany had laid Napoleon III's France low after its victories over Austria and Denmark. Germany was expanding both physically and commercially and Heyde expanded with its country.

Flats were still very popular in Germany itself, so the only real room for expansion lay in export and this Heyde did. Not only did it offer its foreign customers soldiers in any uniform and nationality they wanted, but it also offered large sets depicting such universally known historical episodes as "The Sack of Troy" and "The Triumph of Germanicus". As time went by, it also covered topical subjects: the Zulu War, the Boer War, Arctic exploration, the Balkan Wars.

Although Heyde successfully made parades, it is in the field that its armies are the most impressive. They engage in every conceivable activity. Not only do they fight, but they eat, drink, clean, tend horses, read newspapers, play cards, in other words, as Henry Kurtz has expressed it in *The Art of the Toy Soldier*, "Heyde figures did everything but walk and talk." Even that statement might have to be modified if one considers some of Heyde's standing bands where the conductor leads from a bandstand which is actually a music box. The appropriate national anthem may be played as the clockwork simultaneously moves the bandmaster's baton arm up and down. So, although they still don't talk or walk, they do indeed move and make sounds!

Heyde is also especially famous for its little scenes where several figures are soldered to a tin plate bottom or figures are combined in such a way they easily stand without the help of a footplate. The latter is exemplified by the figure of a wounded man on a stretcher with two attendants examining him. The former might depict hand-to-hand combat or, more often, camp scenes with, for instance, staff officers assembled around a tree-shaded table or cooking fire with attendants. Heyde also made a number of horsedrawn vehicles. One such is an artillery observers' platform on a folding ladder. As far as I know this is a unique piece among lead soldier manufacturers. Heyde's example of the English coronation coach drawn by eight horses is among the most beautiful examples of that oft-depicted coach.

Heyde sold its soldiers in sets contained in square, flat, maroon-colored cardboard boxes. The label was decorated with

*Heyde: combined
medical figures;
value $100 each.*

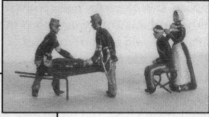

*Heyde: camp scene;
value $150.*

*Heyde: engineers
building a bridge;
value $175.*

*Heyde: boy scouts
lowering a bridge;
value $175.*

medals the company had received and the content described either in handwriting or with a rubber stamp. Large sets were sometimes sold in wooden boxes.

Heyde made soldiers in a great variety of sizes (see table on page 47) of which size 2 (52 millimeters or 2⅟₁₆ inches) is by far the most common. The larger figures, especially horses, were often hollow cast (a process well-known in Germany prior to William Britain's constant use of this manufacturing method, see page 90). Animals, such as elephants, obviously created a problem by their weight so these were also often hollow cast or actually made of a totally different material, composition (see page 236). Ancillary figures such as trees, ruins, fires, explosions, etc. were often semi-flat although tin-plate buildings were also made. Trenches and hills were made of papier maché.

Heyde obtained its many different figures by using plug-in heads, lead which was soft enough to allow bending limbs into appropriate (or, sometimes, inappropriate) positions, the soldering on of equipment, and, of course, by painting the figures in the appropriate uniforms for the regiment or nation's troops depicted. The drawbacks of this system are relatively few, the worst being a tendency for the paint on the arms and legs to peel off and frequent loss of the plug-in heads.

Heyde made a number of personality figures, both German and foreign. Most of these are in the larger sizes, but some were also made in the smaller sizes, especially of royalty. The heroes of the time, such as von Moltke and Bismarck in Germany, Lord Kitchener and Townsend for the English market, as well as some royal figures, are among the most beautiful and sought-after of Heyde's figures.

Heyde made their soldiers in a number of different sizes, the most popular being size 2. The table gives the different sizes for both infantry and cavalry, as given by Heyde. In actuality the figures are somewhat smaller. Thus size 2 standing is closer to 48 (1⅜ inches) than 52 millimeters (2⅟₁₆ inches).

Size	Infantrymen	Inches	Cavalrymen	Inches
3	43	1³⁄₁₆	48	1¼
2	52	2¹⁄₁₆	65	2⁹⁄₁₆
1	58	2⁵⁄₁₆	78	3⅛
00	68	2¹¹⁄₁₆	90	3⁹⁄₁₆
0	75	2¹⁵⁄₁₆	115	4¹¹⁄₁₆
0^{I}	87	3⁷⁄₁₆	120	4¾
0^{II}	120	4¾	145	5¾

Identification   Telling Heyde figures from those made by its large number of competitors is often difficult to impossible for all but the most experienced and knowledgeable specialist. This is why the term "Heyde figure" has almost become synonymous with German solid-cast figures. Only when in their original boxes, which is not common, can a given set easily be ascribed to a given manufacturer. The horses can sometimes be helpful in identification. Heyde horses on the trot were usually semi-solid rather than fully round, whereas their galloping and standing horses are fully round. All the smaller size horses are equipped with a hole for the rider's peg, whereas larger figures sat on separate saddles and horse blankets without the help of the peg.

Prices   Heyde figures are usually found in small groups, parts of sets, some of whose members have disappeared, been bent helplessly out of shape underfoot, or have lost heads and weapons. Sometimes one is lucky enough to come across complete or almost complete sets whose owners were exceptionally good to their toys. Perhaps the owner's parents had paid so much for a set that playing with it was allowed only under the strictest supervision. Rarely does one find complete sets in their boxes. The reason is fairly simple. Heyde figures were toys which were very playable. Parades were what Britains and Mignots were meant for, but Heydes engaged in active service—and suffered far greater casualties for it. Decapitations, loss of limb, lost colors, and worn-away weapons and equipment destroyed were the likely fate of a Heyde soldier, much to the regret of today's collectors.

Single Heyde figures in good condition missing just a little paint and otherwise intact will, in the standard 52 millimeters (2⅛₆ inches) size 2 command a price of $5 to $25 depending on its subject. Marching figures tend to be slightly cheaper than action figures. Mounted figures in a similar condition sell for $15 to $35 and horsedrawn vehicles for $100 to $300. Buying groups is usually a little cheaper, whereas boxed sets are more expensive than the sum of the pieces. Complex figures are more expensive, varying in price from $50 for a simple combination to several hundred dollars for more complicated scenes with several individual pieces combined.

Examples of prices for complete boxed sets of size 2 and, in a separate section, for unboxed complete or almost complete sets are given on the following pages as are the prices for a few other sizes.

Heyde: nurse applying first-aid to Danish soldiers; value $125.

Heyde: artillery observation wagon with ladder extended; value $400.

Heyde: Royal Danish
Guardsman enjoying
his quaff; value $75.

Heyde: artillery obser-
vation wagon with
ladder folded, ready to
move; value $400.

Heyde: English coronation coach from old Heyde catalog;
value $750.

Heyde: King George V, 60 millimeters in size; value $150.
Photograph courtesy of Henry Kurtz Ltd.

Heyde: 80-millimeter
French poilu. Flag
replacement; value
$50 each.

Heyde: 70-millimeter
U.S. mounted bugler;
value $125.

Heyde: 60-millimeter
German soldiers;
value $30 each.

Heyde: 60-millimeter
mounted Indian; value $75.

Heyde: boxed set of dragoons; value $200.

Heyde: boxed set of small size Indians (one horse missing); value $75.

Heyde: Danish Artillery officer
on standing horse; value $25.

Heyde: German lancer on
walking horse; value $25.

Heyde: Danish dragoon on
galloping horse; value $25.

Heyde: Size 1 British lancer;
value $50.

Heyde: Size 1 British hussar;
value $50.

Heyde: crusader;
value $20.

Heyde: bandsmen of the Royal Danish Guard;
value $15 each.

Heyde: engineers at work; value $20 each.

Heyde: Roman legionaries; value $15 each.

Heyde: mounted Roman; value $20.

Heyde: Romans with bear and dog; value $45 each.

Heyde: small Roman chariot; value $125.

Heyde: large Roman chariot; value $200.

Heyde: hollow-cast war elephant with Romans; value $400.

Heyde: British camel corps; value $40 each.

Heyde: Arab on camel; value $75.

Heyde: marching soldiers from different armies; value $15 each.

Heyde: Indian value $10.

Heyde: soldiers from different armies; value $15 each.

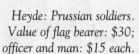

Heyde: Prussian soldiers. Value of flag bearer: $30; officer and man: $15 each.

Heyde: Sailors
in action; value
$15 each.

Heyde: North
Africans attacking.
Value of flag bearer:
$35; men $20 each.

Heyde: British
colonial troops in
action; value
$15 each.

Heyde: British
colonial troops in
action. Value of
man $15; flag is
$30 and mounted
officer is $25.

Heyde: dog sled from polar expedition set; value $100.

Heyde: figures from polar expedition set; value $20 each.

Heyde: figures from polar expedition set; value $20 each.

Heyde Price List

Size 3 (43 mm, 1³⁄₁₆ inches) Boxed Sets

Prussian Lancers (6 pieces)	$400
Italian Infantry (20 pieces)	$200

Size 3 (43 mm, 1³⁄₁₆ inches) Equipment

Horsedrawn cannon	$90
Horsedrawn ambulance	$120
Horsedrawn pontoon wagon	$125
Horsedrawn supply wagon	$90

Size 2 (52 mm, 2¹⁄₁₆ inches)—Complex Figures

Infantry (3 pieces) with cooking fire and tree	$300
Infantry (2 pieces) with cooking fire	$200
Engineers (2 pieces) making bridge	$350
Infantry (2 pieces) in bayonet fight	$150
Infantrist with large radio	$250
Officers (3 pieces) at map table with tree	$450

Size 2 (52 mm, 2¹⁄₁₆ inches) Boxed Sets

Prussian Death Head Hussars in review (12 pieces)	$770
Prussian Guards in review (12 pieces)	$275
Romans, five marching, three mounted (8 pieces)	$440
Band of the Imperial Prussian Guard (20 pieces)	$1,300
Imperial Prussian Guard in review (23 pieces)	$900
Prussian Cuirassiers in review (12 pieces)	$800
Bavarian Infantry in review (25 pieces)	$600
Saxon Infantry of the time in review (23 pieces)	$700
Prussian Cavalry Band (12 pieces)	$500
Medieval Knights mounted and on foot (16 pieces)	$250
German World War I Cavalry (12 pieces)	$600
British Army Signal unit (25 pieces)	$500
American Sailors in action (20 pieces)	$650
7th New York National Guard marching (19 pieces)	$225
Roman Infantry and Cavalry (16 pieces)	$275

U.S. Cavalry (12 pieces) $225
Arctic Exploration set (40 pieces) $3,200

Size 2 (52 mm, 2¹⁄₁₆")—Unboxed Sets or Groups
Indian Army Camel Corps (12 pieces) $1,540
Prussian Infantry 1914 with mounted Uhlans
   (25 pieces) $880
Triumph of Germanicus (38 pieces) $1,200
Ethiopian Campaign 1935 with
   Emperor Haile Selassie (23 pieces) $660
Chicago Police (10 pieces) $300
11th British Hussars in review (24 pieces) $700
Prussian Infantry in action (54 pieces) $600
French Line Infantry in action (42 pieces) $550
German Sailors attacking (9 pieces) $220
British Army encampment (50 pieces) $1,100
U.S. Cavalry in review (12 pieces) $375
U.S. Navy work party (10 pieces) $330
Italian Infantry marching (21 pieces) $350
Knights mounted and on foot (15 pieces) $150
British Lancers and Guards in review (17 pieces) $450
British Guard in review with band (22 pieces) $800
Life Guard with band (12 pieces) $300
French Line Infantry in review with band
   (29 pieces) $600
Napoleon and his staff (mounted) (12 pieces) $600
Russian Infantry in review (17 pieces) $350
Prussian Garde du Corps (7 pieces) $200
American Policemen (24 pieces) $350
West Point Cadets on review with band
   (36 pieces) $1,000
British World War I in action (21 pieces) $350
French Infantry in action (24 pieces) $350
French World War I Infantry in action
   (25 pieces) $400

Prussian Hussars attacking (13 pieces)	$450
Gordon Highlanders in review (14 pieces)	$250
French Zouaves attacking and in review (54 pieces)	$750
French Turkos marching (10 pieces)	$200
Prussian Death Head Hussars attacking (12 pieces)	$400
Boers attacking British Fort (29 pieces)	$900
Mounted Arabs attacking (9 pieces)	$400
Knights on Foot (12 pieces)	$150
French Cuirassiers (7 pieces)	$200
Bedouin Caravan (14 pieces)	$400
Bedouin encampment (9 pieces)	$250
Ethiopians attacking	$250

Size 2 (52 mm, 2¹⁄₁₆ inches)—Equipment

French Horsedrawn artillery set (8 pieces)	$300
Royal Horse artillery set (8 pieces)	$350
French Engineers, Pontoon wagon (8 pieces)	$300
Horsedrawn artillery set (8 pieces)	$250
Hay wagon with two soldiers	$200
Horsedrawn artillery platform	$750

Size 1 (58 mm, 2⁵⁄₁₆ inches)—Boxed

Lord Roberts	$275
Lord Kitchener	$250
King George V	$220

Size OO (68 mm, 2¹¹⁄₁₆ inches) Sets

British Life Guards (12 pieces)	$750
7th New York National Guard in review (18 pieces)	$650
Kaiser Wilhelm I with General Staff, mounted (boxed)(12 pieces)	$2,850

Size O^{II} (120 mm, 4¾ inches)

American Cavalryman (boxed) (1 piece)	$500

♣ GEBRÜDER HEINRICH, MARKE NORIS

This company, the history of which is not well known, had its greatest production at the turn of the century. The output was quite large.

Identification   The figures are very similar to Heyde's but the infantryman has an especially large knapsack and the larger mounted officers have especially prominent beards.

Prices   Similar to those for Heyde. Below a few examples of the approximate prices of some boxed and unboxed sets sold in the last few years.

Scottish Highlanders firing, hollow cast
    (12 pieces) $300
Scottish Highlanders kneeling and firing
    (10 pieces) $400
Mounted Band of the Life Guard (boxed)
    (12 pieces) $1,100

Gebrüder Heinrich: 80-millimeter unknown mounted personality figure; value $150.

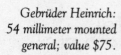

Gebrüder Heinrich: 54 millimeter mounted general; value $75.

Mounted Life Guards (70 mm, 2¹³⁄₁₆ inches)
    (4 pieces) $400
Scottish Highlanders in review with pipe band
    (60 mm, 2⁷⁄₁₆ inches) (26 pieces) $550

✦ JOHANN HAFFNER

This company, which made what was probably the best of all German solid-cast soldiers, was started in 1838 and seems to have lasted for almost one hundred years. At the end of the nineteenth century it was particularly prolific.

    Identification Although the smaller figures are very similar to Heyde's, they are better proportioned with larger, slimmer legs. Groups on lead or tin-plate bases were commonly and very successfully made.

    The boxes are marked J.H. or J.H.N. (Johann Haffner's and Nachfolger, J. H.'s successor). The figures are exceptionally well proportioned and considerably more detailed than Heyde's.

Probably Haffner: British lancer; value $50.

Probably Haffner: British hussar; value $50.

Prices   Because of their higher quality, Haffners, when found and identified, a rare phenomenon, usually command prices 25 to 50 percent higher than the equivalent Heydes.

✤ GEORG SPENKÜCH

Started in 1879 and lasting to 1931 this firm probably came closest to Heyde in its scale of production, both qualitatively and quantitatively. Having started making semi-solids it moved to solids and competed directly with Heyde. In spite of the high quality and great variety of subjects covered, the company went under during the early part of the Depression.

Identification   The boxes of Spenküch are marked G.S., G.Sp., or GSN. The inheritance of the semi-solid era is apparent in many figures which seem to be a cross between the two with (almost) semi-solid bodies and fully rounded heads. Many, however, closely resemble Heyde's but tend to be better proportioned with skinnier limbs and, in some cases, especially well-defined folds in the clothing. They are often on a narrow base with a rounded expansion in the center.

Prices   Similar to Heyde's. Below a few examples of prices obtained in recent years for boxed and unboxed sets are given.

Bavarian infantry in action (24 pieces)	$500
Toy Soldier Target Game (6 pieces)	$500
Roman Infantry and Cavalry (boxed) (18 pieces)	$400

Spenküch: attacking Chinese; value $20 each.

Spenküch: mounted knight; value $20.

Probably Spenküch: elephant (made of hard composition) with mahout; value $100.

Austrian cavalry by Wollner, an Austrian manufacturer of solids; value $30 each.

Roman Battle diorama (24 pieces)	$400
Gauls advancing (22 pieces)	$300

♣ SPANISH SOLIDS

A number of Spanish manufacturers have, over the years, made solid lead soldiers. Starting in 1897, La Guerra of Barcelona put out a 54 millimeter (2⅜ inches) and 45 millimeter (1¹³⁄₁₆ inches) range of mainly Spanish troops.

M. Palomeque started production of its Mignot-like figures in 1922, stopping in 1936 when its factory in Madrid was destroyed during the Spanish Civil War.

From 1940 to 1946 the brothers Sanquez of Madrid produced a series of 50 millimeter (2 inches) figures somewhat similar to Heyde's with plug-in heads. In the early 1950s the company "Multicolor" produced a range of 45 millimeter (1¹³⁄₁₆ inches) Spanish troops in a series on the military history of Spain. These were packaged in small booklike boxes with one mounted or two foot figures.

In recent years the best-known Spanish soldier company has been Alymer (see page 132).

Price for these figures range from $5 to $35, depending on subject and condition.

✤ ITALIAN SOLIDS

The first Italian maker of solid lead soldiers was Figir (Fabbrica Italiana Giocattoli Infrangibili Roma) of Rome, which started production in 1927 and is still in existence. Their series of 54 millimeter (2⅛ inches) solids on oval bases concentrated on the

ISA: Bersaglieri; value $10 each.

ISA: Italian mounted officer (value $20) and infantryman (value $8).

Italian army. The painting is excellent and somewhat reminiscent of Mignot's.

In 1954 the company of ISA (Industria Soprammobili Artistici) of Turin started making 54 millimeter (2⅛ inches) solids. As Figir it concentrates on Italian soldiers but markets a considerable number of sets (three to six pieces) of other armies. Although popular in Italy itself, soldiers of these two producers rarely turn up outside that country.

Prices are in the range of $5 to $35, depending on subject and condition.

✣ BRIGADIER STATUETTE

Carl Martin Andersen of Copenhagen, Denmark started a business of his own in 1946 to fill the void left by Heyde's destruction. A machinist by trade, he founded and led the business "Brigadier Statuette" until his retirement in 1977.

Andersen's figures were of the 50 millimeter (2 inch) size, slightly smaller than 54 millimeter hollow-cast Britains and solid-cast Mignots. Andersen sculpted his figures from a few basic body designs (about twenty) and combining these with about thirty different plug-in heads, he was able to produce a bewildering variety of figures. Mrs. Andersen's painting was meticulous, more so than Heyde or Britain. Often insignias were so detailed, a given figure could be identified as far as regiment and even company.

Andersen, who had been a guardsman himself, produced mainly Danish Royal Guards marching in their red gala uniform. These guardsmen, which were sold in hobby and toy stores in Copenhagen, have always been very popular with Danes and foreign tourists alike. Andersen also made a very wide variety of uniforms, both World War II and older. The Danish army naturally received most attention and almost every Danish uniform from the early nineteenth century to 1945 was made. He also made special figures, on request, for his customers, for instance, the Prussian Guard in parade march. The only horsedrawn vehicle he is supposed to have made was the bridal carriage used at Queen Margrethe II's wedding in 1967. This carriage was exhibited at Thorngren's toy store in Copenhagen for a while (the local equivalent of F.A.O. Schwarz in New York). Lately, however, three Royal Horse Artillery sets made to order have surfaced.

Identification   Although Andersen's main production was of figures on parade, he made many action figures. He did not, however, venture into combined figures as Heyde had. Andersen's horses were, like Heyde's, relatively too small but well proportioned. They came both free standing and on a foot plate. Like Heyde figures, brigader figures are also unmarked. They resemble Heyde's closely but the size is slightly larger (50 millimeters, 2 inches) than the most popular Heyde size 2. The figures of the Royal Danish Guard are easily recognized by their very large bearskin hats with a white cross of light blue on the top.

Prices   Brigadier figures are getting increasingly expensive. Foot figures cost $15 to $25 each and mounted figures $25 to $35. Generally Royal Danish Guard figures are slightly more expensive than, for instance, English soldiers in action.

Brigadier: Royal Danish Guards. Value of flag bearer $35; men $30.

Brigadier: very rare horse artillery; value $400.

Brigadier: German lancer; value $25.

Brigadier: German hussar (Death-head); value $25.

Brigadier: Danish artillery; value $15 each.

Brigadier: German infantry; value $15 each.

Brigadier: soldiers from different armies; value $15 each.

Brigadier: Prussian piper; value $15.

✤ MINIKIN

Although this Japanese company is known to have existed since the 1920s, it was not until the late 1940s that it started producing solid lead soldiers. Many of its figures were mediocre piracies of other toy soldier manufacturers' products, but some, especially of the samurai series, were not only original but of very high artistic quality. Most are of the 54 millimeter (2⅛ inches) scale but some are somewhat larger or smaller. Production stopped around 1960.

Identification Early figures are marked OCCUPIED JAPAN under the base, later ones either MADE IN JAPAN or IMP-JAPAN.

Prices Minikins, especially the figures of the samurai series, are quite attractive and sought after. Most foot figures sell for from $10 to $25 (the latter price for samurai), mounted figures for from $20 to $50. The famous "Historic Hannibal's Elephant Invasion," with a very large and magnificent elephant with attendants, costs $250 to $300.

Minikin: mounted samurai; value $50 each.

Minikin: samurai; value $25 each.

Minikin: soldiers from various armies; value $15 each.

Minikin "The Spirit of 1776"; value $75.

Minikin: mounted figures; value $50 each.

Minikin: copy of Courtenay knight; value $25.

♣ COURTENAY

The figures made by Richard Courtenay (1892–1963) are traditionally considered toy soldiers, but in reality are closer to models than toys. Although Courtenay made other figures, it is for his exquisite knights that he is famous. Early Courtenays are hollow cast but after about 1940 all are solids. They come in a great variety of combat positions (which he numbered), both on foot and mounted. Some consist of two figures on the same stand. Some mounted figures have a movable lance arm. After Courtenay's death in 1963 his molds were taken over by Ping, who reissued them (called Courtenay/Ping) until 1977. After that year the molds were bought by Peter Greenhill, who has reproduced them since.

Identification Although the earliest Courtenays are unmarked, later ones are usually marked with the name of the depicted knight in gold paint on the foot plate and beneath with either MADE IN ENGLAND or MADE IN ENGLAND BY R. COURTENAY. Copies are known to have been made but can generally be told by their far less detailed craftsmanship.

Prices Courtenays are expensive. As they are rarely if ever played with, their condition tends to be excellent. Original Courtenay foot figures sell for $100 to $350, mounted figures for $300 to $500. Courtenay/Ping and Courtenay/Greenhill sell for about half that price.

Courtenay-Greenhill: King Casimir of Poland and Sir John Sarnesfield; value $500 each. Photograph courtesy of Henry Kurtz, Ltd.

Courtenay: Guisard d'Angle, Pier, Sieur de Cramand, Guy, Sieur de Rochefort, Sir John Clinton, and Ardouin de la Touche; value $250 each.

✤ George Grampp & Co.

This small Long Island, New York company produced solid-cast 45 millimeter (1¹³⁄₁₆ inches), 55 millimeter (2³⁄₁₆ inches), and 60 millimeter (2⅜ inches) solid and semi-flat lead soldiers around the time of World War I. It is known that the soldiers were sold in F.A.O. Schwartz's New York store. The figure include U.S. doughboys with Montana hats in turn-of-the-century blue dress uniforms, West Point cadets on parade, English troops with peaked caps, German troops with pickelhue, Poilus with Adrian helmets, Austrian soldiers with caps, and U.S. sailors in both summer and winter uniform, as well as cowboys and Indians.

Recasts from the original molds have occurred irregularly since their discovery in 1972. These recasts have much cruder paint jobs than the originals.

Identification The figures are unmarked. The base is either square or formed as a cross. The heads are cast with the bodies (the somewhat similar Heyde figures from which some may have been copied usually have plug-in heads) and the facial features are characterized by a large nose and rather well-painted details. Mounted soldiers are on prancing horses with a stand under the rear legs.

Prices Foot figures cost about $5 to $10 and mounted figures $10 to $20 in good condition.

Grampp: English and U.S. troops; value $10 each, mounted $15.

Grampp: mounted U.S. officer; value $15.

A selection of Grampp figures; value $10 to $15 each.

✦ OTHER EARLY TWENTIETH-CENTURY UNITED STATES SOLDIER MAKERS

A number of small and generally short-lived companies made solid, hollow-cast and/or semi-round soldiers, cannons, and warships in the period around World War I. Most of their products were of U.S. soldiers and cowboys and Indians and of low quality compared to their European competitors. Not until the advent of the dime-store figures (see page 167) did American toy soldier makers find their own original style. None of these soldiers were marked and their identification is extremely difficult. Their value is low, a few dollars only, although the specialized collectors seek them with enthusiasm.

✤ LINCOLN LOGS

This Chicago-based company, best known for its ingenious interlocking miniature logs, produced a fair amount of solid lead figures. This production started in 1928 and, after an interruption during World War II, continued until the 1950s when the company switched to plastic. The lead figures were sold both in conjunction with the building logs and separately. Some boxes were labeled Noveltoy from a company bought by Lincoln Logs in the mid-1930s. Most were in the 2 to 2⅜ inches (50–60 millimeters) scale but some oversized figures, like Snow White 6 inches (152 millimeters) tall and the seven dwarfs 4 inches (100 millimeters) tall, were also made. They depicted cowboys and Indians, U.S. World War I and revolutionary soldiers, and some farm animals and railroad figures.

Identification   Lincoln Logs figures were marked LINCOLN LOGS USA. Most foot figures are on square bases and, like the unboxed mounted figures, somewhat crude both in casting and painting. The famous "Og, son of Fire" group of prehistoric

Lincoln Logs: Og, Ru, Nada, and Big Tooth; value $50 each.

Lincoln Logs: civilians; value $10 each.

figures are on small platforms with the name of the character in front. The Snow White figures have no base.

Prices   Foot figures fetch $5 to $10 each, mounted $15 to $20. The figures from "Og, Son of Fire" are more sought after and fetch from $15 to $50 each. Even more desirable are the Snow White figures, the dwarfs costing $50 to $100, Snow White herself (fittingly) about $150. Boxed sets cost about 25 percent more than the same of the individual figures.

✣ HISTORICAL MINIATURES

This company produced dime-store size (3 to 3¼ inches—75 to 78 millimeters tall) in solid lead weighing about a quarter pound each. It started production in 1941, was unable to obtain lead during the American participation in World War II, and switched to making the figures in composition (see page 236). It closed down shortly after the war ended. Its short life span and the relatively expensive figures make these rarities.

Historical Miniatures divided their figures into three categories: United States (historical figures, mainly from the Revolution), foreign figures (historical personalities from abroad and a charging highlander), and contemporary figures (both soldiers and personality figures).

A partial list is given on the following page. Known Historical Miniatures figures are listed, as far as possible, in the order used by the company.

United States

George Washington
Benjamin Franklin
John Paul Jones
Lafayette
Steuben
Pulaski
"Spirit of '76" (four figures in a box—son, father, grand-
    father, and Massachusetts regimental flag bearer with
    American flag with stars in a circle)
Officers and soldiers, marching and charging, of Conti-
    nental, Massachusetts, and Pennsylvania regiments[1]
Flag bearers of these regiments[2]
Nathan Hale
Forty-Niner, with genuine gold nugget in pan
Lincoln
Lee
Martha Washington
George Washington mounted

Foreign Figures

Charging Highlander
Simon Bolivar (liberator of South America)
Florence Nightingale (founder of Red Cross)[3]
Baden Powell
Martin Luther
Napoleon

Contemporary Figures

U.S. soldier with flag
U.S. Legionnaire with flag
Winston Churchill

---

[1] Continental: dark blue jacket with orange piping, red vest, beige pants.
Massachusetts: dark blue jacket with white piping, white vest, and pants.
Pennsylvania: brown jacket with white piping, white vest, beige pants.
[2] Massachusetts Regiment flag bearer carries flag with pine tree, the others the
American flag with stars in a circle.
[3] The Red Cross was founded by Henri Dumont, but Historical Miniatures catalogued
her this way.

Black Watch with flag
Greek Evzone with flag
Greek Evzone charging
Cossack in red or green uniform
Franklin D. Roosevelt
General MacArthur
Stalin
Churchill (in naval dress)
General de Gaulle
Chaing Kai-shek
G.I. charging with bayonet (composition)
G.I. charging with bayonet (metal)
Stalin (composition)

Identification It is relatively easy to recognize Historical Miniatures figures. As they are solid cast and large, the weight itself is almost diagnostic (one suggested use was as paperweights!).

The figures are marked with a rubber stamp MADE IN USA (white), or MADE IN USA (black), or in one known case, with a dark blue paper label marked only USA.

Prices Because of their rarity and attractiveness, Historical Miniatures figures are relatively expensive, in mint condition costing between $50 and $75 each.

*Historical Miniatures:
Joseph Stalin;
value $75.*

♣ THE WARREN LINES

Started in 1935 by John Warren Jr., this company produced the most distinguished American solids (some horses were hollow cast). The line was exclusively American including contemporary or World War I cavalry, light field artillery and infantry, all in khaki. Warren used plug-in heads so his soldiers could be equipped with peaked caps or steel helmets. Most of Warren's soldiers have two movable arms. They are 2⅜ inches (60 millimeters), slightly larger than Britains. The company only stayed in business until 1939, partly at least because of the high price of its products.

Warren also produced a contemporary car (a conversion of a Kenton Toy car) used either as a staff car or scout car. All told, Warren produced over 150 figures and used a total of ten different horses. Deservedly, Warren's hollow-cast horses are especially valued for their realistic poses.

Identification Although Warren figures are not marked, they are readily recognized by their size and two movable arms.

Prices The combination of Warren's rarity and high quality makes them expensive. Cavalry figures cost between $100 and $200 each, foot figures $30 to $60 each, the army scout car $500 to $700, and the most expensive the horse drawn field artillery set about $1,000.

Warren: mounted soldier; value $150.

*Warren: scout car with box (soldiers not original); value $600.
Photograph courtesy of Henry Kurtz, Ltd.*

✤ COMET—AUTHENTICAST—S.A.E.
(SWEDISH AFRICAN ENGINEERS)

This sequence of successive toy soldier makers span the period from 1940 to 1970 and three continents.

Comet, a New York die-casting company, began producing 2⅛ inches (54 millimeters) solid-cast soldiers in 1940. These were mainly somewhat crude copies of Britains with a movable arm. The bases were characteristic, oval with square protrusions in front and back. The soldiers were issued under the name Brigadiers in boxes of six to seven. With the outbreak of World War II, Comet switched to producing recognition models for the U.S. government. After the war the company reinstituted its soldier production but now under the name Authenticast. The soldiers were completely redesigned by the Swedish designer Holger Eriksson (1899–1988). These new figures were far more original and well designed, actually lively, than were their Comet predecessors. The solid-cast 2⅛ inches (54 millimeters) scale was maintained but the base was now either cruciform or square. The casting originally took place in the United States while the painting was done in Ireland. The casting was soon moved to Ireland as well, but production difficulties plagued the venture from the start. In 1950 a fire at the Irish factory ended Authenticast but two entrepreneurs from the company, Curt

Wennberg and Fred Winkler, obtained some of the extant molds and moved to South Africa. Here they reissued some of the Authenticast figures under the name S.A.E. (Swedish African Engineers) and Ericksson and others designed new figures for them.

Besides the 2⅛ inches (54 millimeters), 1³⁄₁₆ inches (30 millimeters) scale figures were also made. Both casting and painting was inferior to those of Authenticast but production continued until 1970 although Eriksson had left the company long before.

Identification Comet: The base's shape, oval with small squares protruding in front and back, is diagnostic.

Authenticast: Most are marked EIRE and/or HE (for Holger Eriksson), or F.R. (for Frank Rogers, another designer).

S.A.E.: Most bases are square. Some are marked UNION OF SOUTH AFRICA.

Prices Comet: Single figures sell for about $10 each and boxed sets for $75 to $100. A list of some of the many sets are given below.

Authenticast: Single foot figures sell for $5 to $10 each, boxed sets for $50 to $75 each. Mounted figures are more desirable, selling for $25 to $40, boxed sets for $100 to $150. A representative list of the many sets is given below.

Comet: set of Confederates; value $50.

S.A.E.: Less well made than Authenticast, the 54 millimeter (2⅛ inches) figures sell for slightly less than the Authenticast version. Figures of the 30 millimeter (1³⁄₁₆ inches) scale are not much sought after and sell for less than two dollars each. Boxed sets sell for $20 to $40 each.

Comet: various figures, mainly Britains copies; value $10 each.

Comet: variations on a theme; value $10 each.

Comet: French machinegun team; value $40.

Authenticast box.

Authenticast: Napoleonic artillery; value $125.

Authenticast: mounted soldiers; value $35 each.

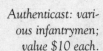

Authenticast: various infantrymen; value $10 each.

Authenticast: various infantrymen; value $10 each.

Authenticast: variations on a theme; value $10 each.

Authenticast: more variations on a theme; value $10 each.

Authenticast: figures of different armies; value $10 each.

Holger Eriksson: Swedish cavalryman in winter dress; value $40.

Holger Eriksson: Swedish soldiers; value $15 each.

Holger Eriksson: Swedish cavalry; value $40 each.

SAE: mounted soldiers. Note the beginnings of lead rot on the Union officer; value $25 each.

SAE: German machinegun team; value $10 each.

SAE: Greek evzones; value $10 each.

SAE: Russians in action; value $10 each.

SAE: 30-millimeter scale U.S. cavalry; value $50.

SAE: 30-millimeter scale U.S. Revolutionary artillery; value $50.

SAE: 30-millimeter scale French cuirassiaers; value $20.

Comet—Samples of the many different sets made

U.S. West Point Cadets marching
U.S. Marines marching
U.S. Sailors marching
U.S Doughboys marching
U.S. Doughboys crawling
Confederates marching
Union troops marching
Colonial Infantry marching
U.S. Marine Band
U.S. Army Band
German Infantry marching
German Infantry attacking
German Alpenjäger
French Infantry marching
French Infantry attacking
Zouaves attacking
French Foreign Legion attacking
Spanish Infantry attacking
English Guards marching
English Infantry marching
English Navy marching
Black Watch marching
Indian troops marching
Greek Evzones marching

Danish Guard marching
Arabs attacking with swords
Knights in Armor standing

Authenticast *(usually sets of seven foot)*
Revolutionary Maryland Regiment
Revolutionary Georgia Regiment
Union Infantry marching
Union Infantry attacking
Confederate Infantry marching
Confederate Infantry attacking
U.S. World War I Infantry marching
U.S. World War I Infantry attacking
U.S. World War II Infantry marching
U.S. World War II Infantry attacking
U.S. World War II Infantry lying shooting
U.S. World War II Military Police marching
U.S. Sailors marching
British Infantry marching (various uniforms)
British Desert Rats marching
British Infantry standing shooting
British Guards marching
Scottish Infantry at ease
Scottish Infantry marching
French Infantry marching (various uniforms)
French Zouaves
French Turcos
French Alpine troops marching
German Infantry marching
German Infantry shooting
German Afrika Korps marching
Russian Infantry marching
Russian Infantry attacking (various uniforms)
Swedish Musketeers
Swedish Guards marching

HOLLOW CASTS

Casting lead figures in such a way that they are left hollow had been done for a long time for larger objects in Germany, but applying the process to the relatively small toy soldiers did not occur until the 1890s.

In about 1890 William Britain, Jr., (1860–1933) started experimenting with hollow casting of lead figures. In this process lead is poured into the mold, but rather than allowing it to completely solidify in the mold, the still-fluid lead in the center is poured out through a hole in the mold as soon as the first thin outer layer has solidified. The mold is then opened and the shiny hollow cast removed and cleaned, at which time the figure is ready for painting. This process takes about ten seconds. It has the advantages of saving a lot of the lead alloy, reducing transportation costs by making the figure light, and still producing great detail on the casting (often later obscured by the painting unfortunately).

Hollow-cast soldiers are easily told from their solid-cast cousins by their weight. It is usually possible to find a small air hole on the top of the head. This hole is sometimes plugged up by lead or paint.

✤ BRITAINS LTD.

William Britain, Sr., (1828–1906) founded what was to become the world's foremost company in the production of toy soldiers around 1845. During the first fifty years the company produced rather clever mechanical toys, but remained small.

Britains' first set of hollow-cast soldiers, five mounted Life Guards, was issued in 1893 and was soon followed by many more. Shortly after Britains' introduction of its hollow-cast soldiers the company instituted and won several lawsuits against competitors who copied Britains soldiers. Subsequently, starting in 1900, paper labels were affixed to the underside of the base of marching soldiers. The same year copyright information (including date of copyright) was engraved on the underside of the horse of the mounted figures and in 1904 on the bottom of the base of the foot figures.

After William Britain, Sr.'s, death in 1906 the company was transformed from a proprietorship under the name of William Britain and Sons to the corporation Britains Ltd. In 1905 a branch was formed in Paris, France. This branch produced the same figures as Britains in England except the engraved MADE IN ENGLAND was substituted by DEPOSE (which means "Registered"). When the Paris office closed in 1923 the molds were sent to England where they continued to be used until worn out. Thus Britains figures with the word DEPOSE engraved may or may not have been made in France. The mold, however, was made there. After a new copyright law was introduced in England in 1911, the date of the copyright was no longer required on the figures and such dates were no longer included in the engraved information on new figures or on new molds. A date on a Britains figure thus indicates only that the mold was first made between that time and 1912 but the figure might be much younger.

During the first World War Britains continued a limited production of its soldiers (the concept of "total war" was not fully developed). After the war, in the twenties, a shift towards civilian figures occurred because of the world's war weariness. Farm and zoo figures as well as cleverly designed garden figures were introduced. With the onset of the depression in the 1930s, Britains tried to introduce cheaper figures, which never really caught on (see below). At the same time, especially as rearmament got under way, the traditional line was expanded to include many more modern sets both of British and U.S. troops (British privates were depicted with black shoes, U.S. soldiers with brown shoes). Equipment, motor vehicles, and airplanes were also introduced in this period, paralleling the transition to motorized transport by the British Army, which was the only major army to enter World War II completely motorized.

During World War II production of lead soldiers stopped completely and war-related items were produced by Britains. Shortly after the end of the war in 1945 Britains again took up the production of lead soldiers, though on a limited scale and initially only for export. New figures were being continually added, the coronation of Queen Elizabeth II in 1953 occasioning a little surge in new models. In 1958 Britains introduced its Farm Picture Packs. These contained one to four figures. Similar Zoo Packs were introduced the following year. Military Picture Packs including some unique figures were introduced in 1954. So-called "Half Boxes," containing only half the number of regular sets, were introduced in 1957. Both these and the Military Picture Packs were discontinued in 1959–1960. At that time many of the regular sets were reduced from eight to seven and mounted figure sets from five to four, called "short sets." A new numbering system (the 9000 series) was introduced in 1962 but production of lead soldiers finally ceased in 1966, a few cannons lasting somewhat longer. Plastic figures, which Britains had started making in 1954, had taken over (see page 277, "Plastic Soldiers"). Seventeen years later, though, Britains seemed to have realized the importance of the collectors market (as opposed to the true toy market) and new metal soldiers were introduced (see page 196, New old toy soldiers).

Identification   Britains foot soldiers of the main series are of the 54 millimeters (2⅛ inch) scale. Most are on square bases. Pre-1907 figures are on round or oval bases. These were

Britains: farm Picture Pack.

Britains: military
Picture Pack.

Britains: #133
Officer from
pre–1907 mold
(round base) and
post–1907 mold
(square base).

maintained only in very few cases (for instance doctors of the medical set #137 and Japanese of set #134). After 1904 copyright information was engraved on the underside of the base of foot figures and on the belly of mounted figures, almost all of which are made in one casting. This information including the words W. BRITAINS LTD., COPYRIGHT, and MADE IN ENGLAND and, more rarely, PROPERTIES, DEPOSE as well as a date (1900 to 1912).

In 1900 Britains introduced one movable arm on most of its figures though not on firing figures. Occasionally both arms are movable but are then usually connected with each other (for instance standing and kneeling officers with binoculars).

The painting is detailed including such things as belts, buttons, and piping. The face usually has highlighted cheeks and, until about 1937, moustaches. Infantry figures wore gaiters until about 1933 when these were replaced with full-length pants.

Most of Britains soldiers were sold in boxed sets, eight foot soldiers or five mounted, in 1960 reduced to seven and four, respectively. Most sets include an officer. Various larger sets combining several figures from other sets were also sold, the largest being set #131 with a total of 275 figures. Most boxes are red with a printed label. This label was initially simple with the name of the regiment and a facsimile of William Britain's signature. About 1906 the artist Fred Whisstock was engaged to make illustrated labels. His labels, which he usually signed, include regimental battle honors as well as (usually) a line drawing of the soldiers of the regiment. Whisstock worked for Britains until 1928 and in the 1930s other labels, also with line drawings, appeared. These carried such texts as TYPES OF THE ROYAL NAVY or ARMIES OF THE WORLD. In 1949 a new four-color label was introduced. This label on the foot soldier boxes depicted a Coldstream Guard and Black Watch Highlands presenting arms, a dismounted Royal Scots Grey Trooper, and a dismounted Life Guard trumpeter on the cavalry boxes. It carried the text "Regiments of All Nations" and these boxes are known as ROAN boxes. In 1961 boxes with a cellophane-covered facing were introduced. Although these boxes had the advantage of showing their content, they were flimsily made. On the end of the box a label was attached showing the number and title of the content. In the period from the late 1950s to the introduction of the cellophane boxes this information was imprinted in white letters directly on the box lid.

When Britains first introduced its motor vehicles in the early 1930s the tires were white and smooth. Later these white tires became ribbed only to be replaced by black ribbed tires immediately prewar. In about 1958 the black rubber tires were replaced with black plastic tires.

Britains cannons and horsedrawn vehicles were initially not painted but had a so-called gun-metal grey finish until the early 1930s, after which various shades of khaki and green were used, as they were on motor vehicles. Very late cannons have a

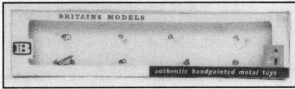

Britains boxes: early simple box, two Whisstock boxes, Types box, ROAN box, and last box with cellophane cover.

Britains: end labels.

very cold green color. Immediately after World War II some cannons were painted dark blue and light green.

Other Britains Besides the main line of 54 millimeters (2⅛ inches) size figures with first grade painting, Britains produced several other lines of toy soldiers.

The A series were second grade figures without movable arms which were produced from about 1918 to 1941. These were for sale in boxes. The C series also consisted of standard size of second grade painting which were for sale in bulk from about 1928 to 1934. This series was continued as the N series until 1940.

The P series also consisted of figures from the same molds as the C series but even more simply painted (bases for instance had the same color as the infantryman's pants). These figures which had been introduced in the early 1930s were reissued in the so-called Crown Range after the Second World War.

Smaller figures with movable arms and first grade painting were issued from 1896 to 1916 in sets. These figures were 43 to 45 millimeters (1¾ inches) tall. Initially they were titled the B series but from 1920 to 1940, the W series. The D series were also 45 millimeter size but with second grade painting. The larger figures (70 millimeters or 2¾ inches) designated the H series of second grade painting were issued in the 1920s and 1930s. They were of a marching Fusilier, line infantrist, and Highlander. Three similar figures of even larger size (83 millimeters or 3¼ inches) were issued in the later 1930s. These were called the HH series.

In the 1950s a series of miniature farm, railroad and hunt figures, as well as both civilian and military vehicles, was sold by Britains, either in bulk or in boxed sets. These figures were meant for the OO and HO trains and were called the Lilliput series. A standing man was 20 millimeters (1⅙ inches) tall. Although military vehicles were made in this size, no soldiers were made.

Britain Regiments are often difficult to distinguish, but their identification is, of course, important if a given figure or set is to be correctly identified.

- Foot Guards in bearskin are recognized by the color (or absence) of the plume on the bearskin cap.
- Scots Guards have no plume.
- Coldstream Guards have a red plume on the right.
- Grenadier Guards have a white plume on the left.
- Irish Guards have a light blue plume on the right.
- Welsh Guards have a white plume with a green stripe on the left.
- The Canadian Foot Guard has a red plume on the left.

Scottish regiments are most easily identified by the pattern of their kilts:

- Black Watch have dark green kilts.
- Argyll and Sutherlanders have dark green kilts with light green crosshatching.

- Gordons have dark green kilts with yellow crosshatching.
- Seaforth have dark green kilts with red and white crosshatching.
- Camerons have dark blue kilts with red and yellow crosshatching.
- Cape Town Highlanders (South Africa) have dark green kilts with yellow crosshatching (like Gordons) but tan jackets.

For detailed description of postwar sets see Joe Wallis, *Regiments of All Nations, Britains Ltd. Lead Soldiers 1946–66.*

For an exhaustive description of prewar Britains until 1932 see Jame Opie, *Britains Toy Soldiers 1893–1932.*

Joe Wallis, *Armies of the World. Britains Ltd. Lead Soldiers 1925–41* covers the middle years.

Britains: second-grade figures.

Britains: second-grade figures.

Britains: second-grade cavalry.

Britains: second-grade cavalry.

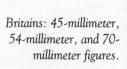

Britains: B series cavalrists.

Britains: 45-millimeter, 54-millimeter, and 70-millimeter figures.

Britains: 70-millimeter figures.

Prices Britains sets are among the most popular toy soldiers and often appear at auctions. As well-advertised auctions almost exclusively with Britains sets take place several times annually both in the United States and in England, prices are very well established. The most desirable are sets in their original boxes and Whisstock boxes are considered especially attractive. Unboxed sets in good to excellent condition are also sought after. It is important that the set is complete and that the individual soldiers are similarly painted. Often collectors or merchants will combine figures from incomplete sets to make a complete set. These mismatched sets are most readily detected by noticing the color of, for instance, the bases (usually green, but quite variable in shade among otherwise similar sets) or, in the case of cavalry, difference in the shade of brown on the horses. Most cavalry sets consist of four mounted men, often two on black horses, two on brown horses, and an officer or bugler on a brown or white horse. The officer's horse, if brown, is not infrequently of a different shade than that of the men's brown horses.

Prices for sets vary tremendously from less than a hundred dollars to several thousands (see sample list below), dependent on the rarity of the set. Sets that were made from the very beginning to the end of production (set #1, the Life Guard) may also vary considerably in price. The earlier the version the more valuable it usually is. The condition of the soldiers is also an important factor in determining the value of a set. Thus even a rare set might, if in poor condition and especially if parts such as arms, bayonets, helmet spikes, and, of course, heads are missing, be almost worthless. The same is true of repainted figures. Unfortunately only the highest priced sets from auctions tend to be quoted in the press, leading to the misconception that all Britains are of great value. These highly priced sets are usually very rare, boxed, and in mint condition, and their prices are of no indication as to the value of other Britains sets as figures.

The price of single pieces from a given set is usually less than the appropriate fraction (usually one-eighth for infantry, one-fifth for cavalry) of the complete set and considerably less than of a boxed set. Officers tend to be slightly higher priced than men. Thus, to use set #1, the Life Guard, postwar is worth $150 in mint, boxed condition. However the single postwar trooper is worth about $20, the officer $25 in mint condition and less if scratched or in other ways imperfect.

In the list below the prices for good, but not necessarily mint, unboxed sets are given. Boxed sets in comparable condition and of similar age are usually worth about 20 to 30 percent more. The number given is Britains catalog number. In some sets the number of figures varied. This is indicated in the text by the boldface number on the left of the price listing.

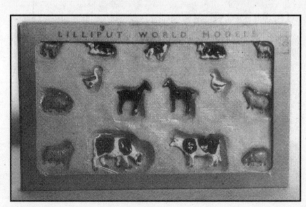

Britains: Lilliput farm series.

*Britains:
Lilliput tank.*

1	Life Guards (5 pieces)	$120
2	Horse Guards (5 pieces)	$120
8	Fourth Hussars (5 pieces)	$120
11	Black Watch charging (6–8 pieces)	$80
17	Somerset Light Infantry (8 pieces)	$100
24	Ninth Lancers (5 pieces)	$130
27	Band of the Line (12 pieces)	$250
28	Mountain Gun of the Royal Artillery (14 pieces)	$350
30	Drums and Bugles of the Line (6–8 pieces)	$100
32	Royal Scots Greys (5 pieces)	$130
33	16th/5th Lancers (5 pieces)	$150

#2 and #8.

*#11 and three
from #17.*

#24 and three from #27.

#28.

#32 and #33.

#35 and two from #36.

35	Royal Marines (8 pieces)	$125
36	Royal Sussex Regiment with mounted officer (7 pieces)	$130
39	Royal Horse Artillery with gun, limber, mounted officer, and 4 outriders (13 pieces)	$450
39A	Same in khaki (13 pieces)	$1,200
44	Queen's Bay (5 pieces)	$200
46	Hodson's Horse (5 pieces)	$200
47	Skinner's Horse (5 pieces)	$175

#39.

#39A.

#44 and #47.

#48.

#74 and two from #75.

*Two from #77
and #82.*

48	Egyptian Camel Corps, 3 riders and 3 camels	$250
66	13th Duke of Connaught's Own Lancers (5 pieces)	$150
67	First Madras Native Infantry at the trail (8 pieces)	$350
74	Royal Welsh Fusilliers with officer and goat mascot (8 pieces)	$180
75	Scots Guard with officer and pipers (7–8 pieces)	$80
76	Middlesex Regiment (8 pieces)	$80
77	Gordon Highlanders with piper (6–8 pieces)	$100

79	Royal Navy landing party (11 pieces)	$300
82	Scots Guard Colours and pioneers with axes (7 pieces)	$175
90	Coldstream Guards firing in 3 positions with 2 officers, bugler, and drummer (30 pieces)	$300
98	Kings Royal Rifle Corps (8 pieces)	$200
101	Band of the Life Guard (12 pieces)	$400
110	Devonshire Regiment (8 pieces)	$275
112	Seaforth Highlanders (8 pieces)	$150

#90.

#98.

#101.

114	Cameron Highlanders (8 pieces)	$200
115	Egyptian Lancers (5 pieces)	$150
117	Egyptian Infantry (8 pieces)	$150
120	Coldstream Guard kneeling firing (8 pieces)	$130
123	Bikamir Camel Corps (3 pieces)	$300
124	Irish Guard lying firing (8 pieces)	$175
133	Russian Infantry (8 pieces)	$200
136	Russian Cavalry (Cossacks) (5 pieces)	$150
134	Japanese Infantry charging (8 pieces)	$400

#110 and #114.

#115.

#117.

#133 and #136.

#134 and #135.

#137.

*#138, #141,
and #142.*

135	Japanese Cavalry (5 pieces)	$500
137	Royal Army Medical Corps (24 pieces)	$450
138	French Cuirassiers (5 pieces)	$135
141	French Infantry of the Line (8 pieces)	$250
142	Zouaves charging (7–8 pieces)	$120
145	Royal Army Medical Corps Horsedrawn Ambulance (7 pieces)	$400
146	Army Service Corps Wagon (5 pieces)	$250
147	Zulus of Africa (8 pieces)	$120
150	North American Indians on foot (8 pieces)	$100

#145.

#146.

#147.

#150.

152	North American Indians on horses (4–5 pieces)	$100
159	Yeoman, Territorial Army (Yeomanry) (5 pieces)	$300
160	Territorial Army Infantry at trail (8 pieces)	$225
162	Boy Scout Camp (23 pieces)	$600
163	Boy Scout signalers (5 pieces)	$350
164	Arabs on Horses (5 pieces)	$130
169	Bersaglieri (8 pieces)	$160
179	Cowboys, 4 mounted, 1 foot (5 pieces)	$100
182	11th Hussars dismounted with horses (8 pieces)	$160
183	Cowboys on foot (7–8 pieces)	$100

#152.

#159 and #160.

#162.

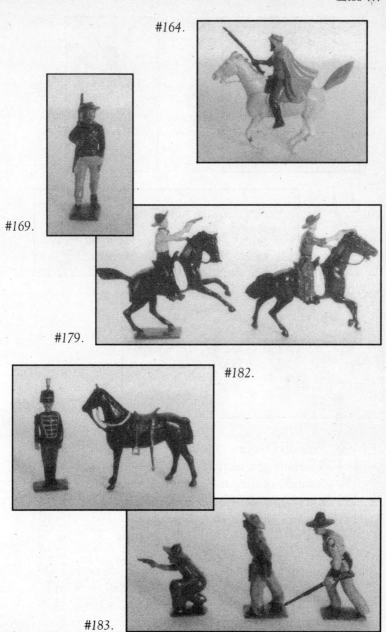

#164.

#169.

#179.

#182.

#183.

#190.

#196.

#194 and
#198.

#200 and #199.

190	Belgian Cavalry (5 pieces)	$200
194	Machine gun section lying (8 pieces)	$100
195	Infantry of the Line at the trail (8 pieces)	$120
196	Greek Evzones (7–8 pieces)	$150
197	Gurhka Rifles at trail (8 pieces)	$150
198	British Machine Gun section sitting (6 pieces)	$120
199	Motorcycle Machine Gun Corps (3 pieces)	$300
200	Dispatch Motorcycle Riders (4 pieces)	$175
201	Officers of the General Staff (4 pieces)	$200
202	Togoland Warriors (8 pieces)	$125

203 Pontoon Section, Royal Engineers (7 pieces)$500
212 Royal Scots marching at the slope
      (5–8 pieces) $150
216 Argentine Infantry (8 pieces) $250
217 Argentine Cavalry (4–5 pieces) $250

#201.

#202.

#212.

#216 and
#217.

224 Arabs of the desert, 2 on camels,
     4 marching, 2 on horses, 1 large,
     2 small palm trees (11 pieces) $350
225 King's Africa Rifles (8 pieces) $150
226 West Point Cadets, winter dress (8 pieces) $125
227 U.S. Infantry at the slope (Doughboys)
     (8 pieces) $100
228 U.S. Marines at the slope (8 pieces) $125
229 U.S. Cavalry (5 pieces) $100
230 U.S. Sailors, blue jackets (8 pieces) $150
258 World War I British Infantry at the trail
     with gasmasks (8 pieces) $90
299 West Point Cadets, summer dress (8 pieces) $100

#225.

#224.

#226, two from
#227, and two
from #228.

#229.

#230.

#258.

#299.

#312.

312	Grenadier Guards in winter coats (8 pieces)	$150
313	Royal Artillery, kneeling and standing (8 pieces)	$200
329	Sentry (Grenadier and Scots Guard) and Sentry Box (2 pieces)	$35
400	Life Guard, winter dress (5 pieces)	$175
432	German Infantry (8 pieces)	$120
435	U.S. Monoplane with pilot and hangar (3 pieces)	$2,750

#313.

#400.

#432.

#460.

460	Color party of the Scots Guard (7 pieces)	$300
1201	Gun of the Royal Artillery (1 piece)	$25
1202	Carden Lloyd Tank with 2 men and machine gun (4 pieces)	$150
1250	Royal Tank Corps (8 pieces)	$400
1251	U.S. Infantry (Doughboys) firing with officer (9 pieces)	$300
1253	U.S. Sailors, white jackets (8 pieces)	$150
1257	Yeoman of the Guards (Beefeaters) (9–11 pieces)	$175

#1201.

#1202 and #1250.

#1251.

#1253.

#1257.

1258 Knights in armor, 3 mounted, 3 on foot
    (6 pieces) $300
1260 British Infantry firing (9–10 pieces) $250
1263 Gun of the Royal Artillery (1 piece) $20
1264 Naval Gun (1 piece) $50
1266 Heavy Howitzer on wheels (1 piece) $100
1283 Grenadier Guards firing (8 pieces) $100

#1258.

#1260.

#1263.

#1264.

1290 Band of the Line in Service dress
(black shoes) (12 pieces) $400
1292 Gun of the Royal Artillery (1 piece) $20
1301 U.S. Band in khaki (brown shoes)
(12 pieces) $400
1307 Sixteenth-century Knights, 6 mounted,
3 on foot (9 pieces) $125
1313 Eastern People, 5 arabs, 2 goats, 1 donkey,
1 camel with young, 2 palm trees (11 pieces) $800

#1290.

#1292.

#1307.

#1313.

1321 Armored car (1 piece) $350
1327 Grenadier Guards firing (14–16 pieces) $150
1330 Royal Engineers Service Wagon (4 pieces) $300
1333 Army Lorry with halftrack and driver
     (2 pieces) $250
1334 Army Lorry with driver (2 pieces) $150
1335 Army Lorry with 6 wheels and driver
     (2 pieces) $200
1343 Royal Horse Guard, winter dress (5 pieces) $250
1349 Royal Canadian Mounted Police (5 pieces) $150
1383 Belgian Infantry firing (14 pieces) $350
1389 Belgian Infantry at the slope (8 pieces) $250
1424 Emperor of Ethiopia's Body Guard (8 pieces) $400
1425 Ethiopian Tribesmen (8 pieces) $175

#1321.

#1330.

#1333.

#1335.

#1349.

Two from
#1383
and 1389.

#1424 and
#1428.

*#1435, #1436,
and #1437.*

#1448.

#1470.

1435	Italian Infantry (8 pieces)	$200
1436	Italian Infantry Colonial dress (8 pieces)	$300
1437	Italian Carabiniere (7–8 pieces)	$200
1448	Staff car with men ("Siamese twins") (2 pieces)	$300
1470	State (Coronation) Coach (11 pieces)	$350
1475	Yeoman of the Guard (Beefeaters) and walking outriders and footmen (19 pieces)	$350
1479	Royal Artillery limber (1 piece)	$100
1512	Army Ambulance with driver, wounded man, and stretcher (4 pieces)	$200

1515 Coldstream Guard at the slope (8 pieces) $100
1518 Line Infantry 1815 (8–9 pieces) $275
1519 Highlanders 1815 (8–9 pieces) $275
1527 Royal Air Force Band (12 pieces) $500
1542 New Zealand Infantry at the slope (8 pieces)$250
1544 Australian Infantry at the slope (8 pieces) $250
1554 Royal Canadian Mounted police on foot
    (7–8 pieces) $150

*Two from #1518
and two from
#1519.*

#1542.

#1554.

1555 Changing of the Guard at Buckingham
Palace (83 pieces, boxed) $1,500
1603 Irish Infantry at the slope (8 pieces) $250
1610 Royal Marines presenting arms (8 pieces) $225
1611 British Infantry in gasmasks, crawling
(8 pieces) $250
1612 British Infantry in gasmasks, throwing
grenades (8 pieces) $100
1612 British Infantry in gasmasks, charging
(6–7 pieces) $100
1631 Governor General (Canadian) Horse
Guard (5 pieces) $150
1633 Princess Patricia's Canadian Light Infantry
(8 pieces) $150

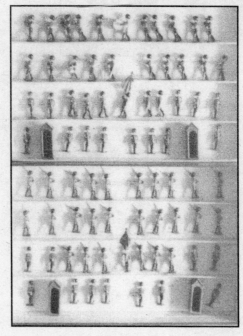

#1555.

#1603.

Two from
#1613 and
#1612.

#1631.

#1633.

#1638.

#1639.

#1641 with #1640 mounted.

#1662.

1634 Govenor General (Canadian) Footguard
　　　at the slope (7–8 pieces)　　　　　　$150
1638 Sound Locator with operator (2 pieces)　　$50
1639 Range finder with operator (2 pieces)　　$35
1640 Searchlight (1 piece)　　　　　　　　$75
1641 Heavy duty underslung lorry and driver
　　　(2 pieces)　　　　　　　　　　　$350
1659 Mounted knight with mace (1 piece)　　$75
1660 Mounted knight with sword (1 piece)　　$75
1661 Mounted knight with lance, charging
　　　(1 piece)　　　　　　　　　　　　$75
1662 Mounted knight, standard (1 piece)　　$75
1663 Mounted knight with lance, rearing (1 piece) $75
1664 Knights on foot (5 pieces)　　　　　$175
1711 Foreign Legion at the slope (6–7 pieces)　$140
1715 2 pounder AA gun (1 piece)　　　　　$25

#1664.

#1711.

1717	Mobile 2 pounder AA gun (1 piece)	$75
1718	Mobile searchlight (1 piece)	$85
1720	Band of the Scots Grey (7 pieces)	$400
1723	Royal Army Medical Corps, 2 stretcher bearers, 1 wounded, 2 nurses (9 pieces)	$150
1725	4½ inch Howitzer (1 piece)	$25
1726	Limber (1 piece)	$25
1728	Predictor with operator (2 pieces)	$30
1729	Height Finder with operator (2 pieces)	$40
1731	Spotting chair with spotter (2 pieces)	$30

#1717.

#1718.

#1720.

#1723.

#1725.

#1726.

#1728.

#1731.

#1856.

#1858.

#1876.

#1877.

#1893.

1856 Polish Infantry at the slope (8 pieces) $350
1858 British Infantry with slung rifles, bren
     gunner, and officer (8 pieces) $125
1876 Bren Gun carrier with crew (4 pieces) $100
1877 Beetle lorry with river and cover (3 pieces) $125
1893 Royal Indian Army Service Corps. with
     mule and officer (7 pieces) $250
1898 British Infantry standing with rifles and
     tommy guns and officer (8 pieces) $75
1901 Capetown Highlanders (8 pieces) $300
1907 British army staff officer with dispatch
     rider (5 pieces) $200

#1898.

#1907.

#1901.

1911 Officers and Petty Officers of the Royal
Navy (7 pieces) $200
2009 Belgian Grenadiers at the slope (8 pieces) $200
2010 Parachute Regiment ("Red Devils")
(7–8 pieces) $175
2017 Ski Troops (4 pieces) $500

#1911.

#2009.

#2010.

#2017.

2018 Danish Guard Hussars (8 pieces) $700
2019 Danish Life Guard (7 pieces) $250
2021 U.S. Military Police ("Snowdrops")
      (8 pieces) $125
2022 Swiss Papal Guard (9-11 pieces) $300
2025 Cameron Highlanders firing with officers
      and pipers (14–18 pieces) $300

#2018.

#2019.

#2022.

#2025.

2026	25 Pound Howitzer (1 piece)	$15
2027	Red Army Guards in coats (8 pieces)	$150
2028	Red Army Cavalry (5 pieces)	$200
2029	Life Guards at the halt, 3 mounted, 3 on foot (6 pieces)	$120
2031	Australian Infantry in battle dress (8 pieces)	$150
2032	Red Army Infantry marching in review (8 pieces)	$150
2033	U.S. Army at the slope (steel helmets) (8 pieces)	$100
2034	Covered Wagon ("Prairie Schooner") with pioneer and wife (7 pieces)	$200
2035	Swedish Life Guard (8 pieces)	$200

#2026.

#2027.

#2031.

#2028 and
#2032.

#2033.

#2034.

#2035.

#2044.

#2051.

#2055.

#2056.

2041 Trailer with clockwork engine (1 piece) $100
2043 Rodeo set (13 pieces) $500
2044 U.S. Air Corps with slung rifles (8 pieces) $130
2051 Uruguayan Military School cadets
    (8 pieces) $300
2055 Confederate Cavalry (5 pieces) $125
2056 Union Cavalry (5 pieces) $125
2057 Union Artillery with 2 gunners (3 pieces) $85

#2057.

#2059.

#2060.

2058 Conference Artillery with 2 gunners
    (3 pieces) $85
2059 Union Infantry (7 pieces) $100
2060 Confederate Infantry (7 pieces) $100
2062 Seaforth Highlanders charging with
    2 pipers and mounted officer (15–17 pieces) $300
2063 Argyll and Sutherlanders firing (5–6 pieces) $125
2064 U.S. 155 millimeter gun with detachable
    rear wheels (2 pieces) $125

2065 Queen Elizabeth II on horseback (1 piece) $60
2067 Sovereign Standard of the Life Guard
     and escort (7 pieces) $300
2071 Royal Marines presenting arms
     (6–7 pieces) $125
2072 King's Royal Rifle Corps at the trail
     (8 pieces) $250
2073 Royal Air Force at the slope (8 pieces) $250
2074 First Dragoon Guards (5 pieces) $200

#2072.

#2073.

#2074.

2075 7th Queen's Own Hussars (5 pieces) $200
2076 12th Lancers (5 pieces) $150
2078 Irish Guards presenting arms (7 pieces) $150
2079 Royal Company of Archers (13 pieces) $450
2080 Royal Navy at the slope (7–8 pieces) $150

#2075.

#2076.

#2079.

#2080.

#2085.

#2086.

#2087.

2082	Coldstream Guard attention (8 pieces)	$150
2083	Welsh Guard at ease with mounted officer (7 pieces)	$175
2084	Color party of the Scots Guard (6 pieces)	$300
2085	Household Cavalry Musical Ride (23 pieces)	$1,200
2086	West Surrey (Queen's) Royal Regiment firing (14-16 pieces)	$250
2087	Fifth Royal Inniskilling Dragoon Guards (8 pieces)	$300

#2088.

#2089.

#2090.

#2091.

2088	Duke of Cornwall Light Infantry (8 pieces)	$300
2089	Gloucestershire Regiment (8 pieces)	$300
2090	Royal Irish Fusiliers (8 pieces)	$300
2091	Rifle Brigade at the trail (8 pieces)	$300
2092	Parachute Regiment at the slope (8 pieces)	$300
2094	State Open Landau (11 pieces)	$300

#2092.

#2094.

#2095.

2095 French Foreign Legion in action
(12-14 pieces) $350
2098 Venezuelan Military School Cadets at
the slope (7 pieces) $300
2101 U.S. Marines Color Guard (4 pieces) $200

#2098.

#2101.

#2102.

#2110.

2102 Austin "Champ" with removable canopy
(2 pieces) $100
2110 U.S. Military Band ("Yellow Jackets")
(25 pieces) $2,000
2148 Fort Henry Guards (Canadian) with
goat mascot (7 pieces) $125

2150 Centurion Tank with detachable antennae
    (3 pieces) $400

2153 United Nations Infantry with slung rifles
    (8 pieces) $500

2173 Battalion Anti-tank gun (1 piece) $50

#2148.

#2150.

#2153.

#2173.

Picture Packs boxed in excellent condition vary from $25 (foot) to $60 (mounted).

Farm figures are about $10–$15 for the small animals, $15–$25 for the larger animals, and $10–$25 for people. Horsedrawn vehicles are priced from $50 to $150.

Hunt series range from $20 (foot) to $30 (mounted) with the dogs being about $10. Set #234 "The Meet" of eighteen pieces and set #235 "Full Cry" of twenty pieces are $400 each.

Picture Pack; value $50.

Farm figures; value $15 each.

Hunt figures. Value: dogs $10 each; mounted figures $30, and foot $20.

Zoo animals;
value $30.

Circus figures;
value $50 each.

Railroad figures;
value $40 each.

Soccer figures; value $30 each.

Zoo animals range from $10 (small animals) to $100 (elephant).

Circus figures are about $40 each. The Mammoth Circus set #1539 of twenty-three pieces is priced at about $1,000.

Railway figures are priced from $20 to $60.

Soccer teams cost about $800, individual players and umpires $25 to $40 each.

Second grade figures should sell for $10-$15 each.

Small size (43 to 44 millimeter or 1¾ inches) Britains (black and white series) vary in price from $10 to $20 for excellent pieces.

✤ JOHILLCO

This company, officially called John Hill and Co., was started at the turn of the century by a former employee of Britains, George Wood, and continued producing hollow-cast 54 millimeter (2⅛ inch) figures until the 1950s. At that time it switched to making plastic figures the output of which, however, was very limited as the company folded in the 1960s.

Although Johillco exported to the United States, this export was quite limited compared to Britains, and they competed with Britains mainly on the home market in England. They met with considerable success in this competition as their figures were cheaper and mostly sold singly. The factory was destroyed during World War II, and although production continued after the war, it never seemed to recover enough to reach prewar levels.

Identification   Most of Johillco's figures are marked, foot figures on the foot plate, mounted on the horse's belly. The markings include variations of the name such as J. HILL & CO., JOHILLCO, JO HILL, J. HILL and some, especially animals, only have ENGLAND or MADE IN ENGLAND on the casting. Although reminiscent of Britains, Johillco figures are often more animated in their positions, sometimes in anatomically hopeless positions, with simpler painting. Some have movable arm(s), but

Johillco: bugler,
drummer, and
advancing
English soldier;
value $10 each.

Johillco: British cavalry, standing and laying firing; value $10 each.

Johillco: British attacking, sitting, and laying with machinegun;
value $10 each.

Johillco: motor-
cycle; value $25.

most do not. Many, especially animals, show a clear casting line on the underside.

Johillco made a large variety of figures depicting British troops as well as American soldiers, knights, cowboys and Indians, Romans, and Zulus. Their civilian line was extensive including both farm and zoo animals.

Prices Most Johillco figures were sold singly and only rarely are boxed sets found. Most of the list below therefore refers to single figures in good, but not mint, condition.

Johillco: Gordon's highlander; value $8.

Johillco: Scots Grey; value $15.

Johillco: Confederate and Union soldiers. All rifles broken; value $5 each.

Johillco: Arab on camel; value $25.

Johillco: Arabs on horses; value $20 each.

Johillco: British camel corps; value $20. Greek evzone; value $10.

Boxed sets

131	71st New York National Guard Regiment marching (7 pieces)	$75
132	7th New York National Guard Regiment marching (7 pieces)	$75
157	Royal Canadian Mounted Police at east with mounted officer (7 pieces)	$75

C7	North American Indians in action with mounted chief (7 pieces)	$75
133	U.S. Marines marching (7 pieces)	$75
127	Greek Evzones marching (7 pieces)	$55
A	Coronation coach (small size)	$65
181	Life Guard mounted and on foot (6 pieces)	$60
105	Life Guard mounted (4 pieces)	$60
122	Scottish Highlanders firing (8 pieces)	$75
D202	West Point Cadets and U.S. Marines in review (14 pieces)	$175
202	Camel Corps (6 pieces)	$150
304	Knights and Crusaders mounted and on foot (17 pieces)	$200
C5	Mounted cowboys	$75

Single figures

33P	Scots Grey mounted	$15
215A	Scots Guard marching	$8
243A	Scots Guard piper	$10
12A	Highlander with tropical helmet lying firing	$10
615A	Prone machine gunner in khaki	$15
13D	Standing drummer in khaki	$10
535A	Cavalry man in khaki	$15
591D	Dispatch rider in khaki on motorcycle	$30
9A	Infantryman attacking in khaki	$10
21C	Sailor kneeling with rifle	$12
?	Confederate soldier lying firing	$10
172L	Mounted knight with lance	$20
?	Standing knight	$10
916A	Kneeling cowboy with revolvers	$8
589A	Mounted cowboy with rifle	$20
20C	Standing Indian with tomahawk	$8
902C	Mounted Indian shooting rifle	$20
755	Stagecoach with 2 horses and 2 men	$100
677	Arab mounted on camel	$30

40P	Arab mounted on horse	$20
177P	Mahmout on elephant	$20
596A	Roman soldier on foot with sword	$10
581	Roman chariot with 2 horses and driver	$100
307	Cow	$5
309	Horse grazing	$5
246 OW	Pig	$5
290	Sheep	$3
276	Rabbit	$5
279	Duck	$3
284	Chicken	$3
258	Farmer	$10
345	Greenhouse	$100
380	Gazebo	$100
278	Pond	$40
246F	Foot bridge	$15
246B	Oak tree	$10
157	Hedge	$7
600W	Penguin	$4
600A	Lion	$12
600R	Camel	$15
600S	Alligator	$12

Johillco: mounted Royal Canadian Mounted Police: value $20; standing Royal Canadian Mounted Police: value $15; West Point cadet: value $10; and Finnish skitrooper: value $25.

*Johillco: knight;
value $15.*

*Johillco:
gladiator;
value $8.*

*Johillco: Horse Guard;
value $20.*

Johillco: dragoons; value $12 each.

✤ TIMPO

The official name Toy Importers Ltd. was contracted to "Timpo," under which name its figures are known.

Timpo started production of hollow-cast 54 millimeter (2⅛ inch) figures immediately after World War II and attained a consistent and very attractive series of figures. They are quite different from Britains, both men and animals being heftier and in very natural poses. Especially noteworthy are their series of knights, GIs, and West Point cadets, all of which attained a measure of popularity in the United States as well as England.

In 1956 production shifted to plastic with a consequent drastic drop in quality. The company ceased production altogether in 1979.

Timpo: English tommy ready for travel; value $30.

Timpo: English police; value $12 each.

Timpo: Life Guard;
value $28.

Timpo: Queen Elizabeth II;
value $45.

Identification   Most, but not all, figures are marked with the word ENGLAND in characteristic block letters across the back, animals under the belly. The figures are, with their hefty stature and excellent painting, quite readily distinguished from other hollow-cast figures. Many knights have ornaments of colored hair on their helmets, a diagnostic feature when present. Horses and mounted men are cast separately.

Prices   Recently more collectors have been showing interest in these attractive figures, with a consequent rise in

Timpo: boxed set of GIs; value $150.

prices. Boxed sets are rather uncommon and most of the prices below refer to individual figures in good, but not mint, condition. Soldiers of the GI series painted as Afro-Americans cost about twice the regular amount.

Timpo: GI marching. Value of flag: $30; others $15.

Timpo: GIs in action; value $15 each.

Timpo: GIs in action; value $15 each.

Boxed sets

950	GIs in action (16 pieces)	$200
908	GIs in action (6 pieces)	$75
902	GIs marching (7 pieces)	$150
B30	British infantry in action (9 pieces)	$150
B12	British infantry in action (6 pieces)	$100

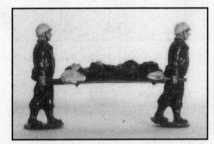

Timpo: GI stretcher team; value $45.

Timpo: Afro-American GIs; value $30 each.

Timpo: West Point cadet musicians; value $15 each.

Timpo: West Point cadet musicians; value $15 each.

Single figures

3000	British Guard Drum Major	$10
3002	British Guard Drummer	$10
3010	British Guard marching	$10
3018	Scottish officer	$15
3019	Scottish marching	$15
3020	Scottish Piper	$18
3017	Royal Horse Guard mounted	$30
8009	Sailor with luggage awaiting train	$35
8010	Soldier with luggage awaiting train	$35
9000	GI at ease	$15
9004	GI crawling	$15
9011	GI standing shooting	$15
9007	GI on motorcycle	$50
9012	GIs (2) with mortar (metal or plastic)	$35
9024	GI stretcher team with wounded	$40
9020	GI marching with forage cap	$15
9021	GI flag bearer with forage cap	$30
7000	West Point Cadet Drum Major	$15
7002	West Point Cadet Drummer	$15
7009	West Point Cadet Flag bearer	$35
7012	West Point Cadet marching	$15

7016	West Point Cadet mounted Officer	$35
7014	West Point Cadet standing firing	$15
KN58	Mounted Knight (Front de Boeuf), Ivanhoe series	$35
HF502	Mounted Duke's General, Quentin Durward series	$45
WW2000	Mounted bandit with 2 revolvers	$15
WW2022	Sheriff standing with revolver	$10
WW2014	Mounted Indian with spear	$18
WW2019	Indian running with tomahawk	$10
WW2012	Cowboy tied to tree	$25
MF1000	Cow standing	$8
MF1004	Shire Horse standing	$8
MF1015	Pig standing	$5
MF1010	Sheep standing	$5
MF1022	Chicken	$3
MF1028	Farmer	$8
MF1031	Milkmaid with pail	$8
MZ4000	Elephant	$30
MZ4002	Camel	$15
MZ4008	Tiger	$10
MZ4025	Pelican	$5

Timpo: West Point cadets; value $15 each.

Timpo: West Point cadets. Value of mounted and flag: $35; others $15 each.

Timpo: Hopalong Cassidy; value $40.

✤ CHARBENS & CO.

Established in 1930, Charbens produced a number of 54 millimeter (2⅛ inch) figures of only medium quality. Their range included not only modern soldiers (the British 8th army and the Africa Korps), but also many knights, pirates, etc.

 Identification Most figures are unmarked, making identification difficult. Their series of American GIs, which is very similar to that of Timpo's, can be recognized by the rimless helmet. Charbens figures do not have movable arms.

 Prices Charbens figures are generally much sought after. Most figures are in the $5 to $15 range, but special sets,

Charbens boxed set of GIs; value $250. Photograph courtesy of Henry Kurtz, Ltd.

such as the seven-piece Jazz Dance Band set ("Jack's Band") have fetched over $500 at auction.

♣ CRESCENT

Crescent Toy Company, Ltd. did not start making toy soldiers until about 1930. Its range of 54 millimeter (2⅛ inch) hollow-cast figures of mediocre quality included modern troops, Zulus, cowboys and Indians, arabs, Royal Canadian Mounted Police, a British Royal Horse Artillery team, and a naval landing party with cannon. In 1956 Crescent turned to plastic.

Crescent: Royal Canadian Mounted Police; value $5 to $10 each.

Crescent: Arab on camel; value $15.

Identification   Most Crescent figures are unmarked. Some Crescent figures are marked with a crescent moon, the company's official logo, and with the name CRESCENT TOYS ENG. Most figures do not have movable arms.

Prices   Crescent figures are not much sought after. Single figures are in the $5 to $15 range. A selection of approximate prices obtained for boxed sets are given below.

2154	Armored Patrol unit	$75
2404	North West Mounted Police (6 pieces)	$30
	Indians in action (8 pieces)	$50

Crescent: hunter on elephant; value $15.

Crescent: U.S. marine; value $8.

Mounted cowboys (7 pieces)	$50
Cowboys and Indians display set (19 pieces)	$125
U.S. Marines in review (14 pieces)	$75

♣ CHERILEA

Started in 1948, Cherilea Products of Blackpool, England, made a number of undistinguished British ceremonial and modern soldiers and a few excellent mounted knights of which the "Black Prince" is the most famous. A set of ten baseball players with the umpire is also noteworthy. In the 1950s the company switched to plastic.

Identification   Cherilea's figures are marked with CHERILEA in script.

Prices   Because of the rather poor quality, most Cherilea figures fall in the $5 to $15 range. Boxed sets, like a five-piece set of mounted Life Guards, fetches $35 to $65.

Cherilea: U.S. marine, sailor, and Air Force Womens Corps.; value $8 each.

Cherilea: Scottish piper and soldier; value $8 each.

*Cherilea: knight;
value $15.*

*Cherilea: knights;
value $8 each.*

✣ EUREKA/AMERICAN SOLDIER COMPANY

This company appears to be the first successful American toy soldier company. Founded in 1898 by Charles W. Beiser, it marketed its hollow-cast soldiers under the name Eureka until 1903. From then on only American Soldier Company was used. In 1904 Beiser patented a tray with hinged clasps which could hold a soldier. This tray made it possible to market the soldiers in boxes with a cork gun with which the soldiers could be shot down as in a shooting gallery. These sets became quite popular. For a period between 1906 and 1915 many of Beiser's sets came with Britains soldiers or a mixture of American-made and Britains soldiers. Britain had obtained permission to use the tray presumably for payment in kind (soldiers). The company folded in the late 1920s.

Identification   Boxed sets are marked with paper labels. The individual soldiers produced by American Soldier Company are unmarked. They are 52 millimeters (2¹⁄₁₆ inches) or 54 millimeters (2⅛ inches) in size, usually hollow-cast figures which look like simplifications of Britains figures of the period. The paint work is quite meticulous, especially of the face. The individual figures are unmarked. The subject is American soldiers of

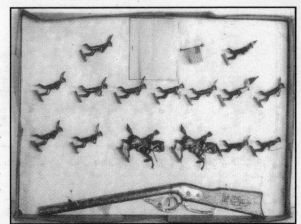

American Soldier Company: Naval set in box with cork gun; value $400.

American Soldier Company: Naval set "ready for action;" value $400.

the time and cowboys and Indians. One subject used Britains Zulus. The sets came in different sizes containing from four to twenty-four figures plus tent(s) and/or cannons as well as the pop gun and instructions on how to play the game.

Prices Individual Eureka American Soldier Company figures are worth $5 to $25. Boxed sets in the original trays sell for from $100 to $1,000, depending on size and completeness of the set.

American Soldier Company: Spanish War figures. Value of mounted $15; others $10 each.

✤ McLoughlin

This New York company is best known for its paper soldiers (see page 153), but also produced a number of hollow-cast lead soldiers of 1⅞ inches (48 millimeter) size, which were sold in colorfully labeled boxes in the early part of the twentieth century.

Identification McLoughlin's lead soldiers are all marching or mounted on parade. The uniforms are the turn-of-the-century U.S. army uniforms in different color variations. The square base is fairly thick with two characteristic indentations on the bottom, but are otherwise unmarked. An airhole is found on the side of the helmet.

McLoughlin U.S. troops. Value of mounted $15; others $10 each.

Prices Single McLoughlin figures sell for $5 to $15, whereas boxed sets of from eleven to thirty-seven pieces sell for $100 to $350, depending on size, condition, and completeness.

♣ DIME-STORE SOLDIERS

The term "dime-store figures" refers to about 3¼ inch (82 millimeter) figures made from the early 1930s to 1950s. These figures were sold in the ubiquitous five- and ten-cent stores of the period, especially Woolworth's. The price for most figures was a nickel, more complex ones a dime. Although soldiers were the main subjects, many others were depicted, farmers, clowns, cowboys, firemen, and so on.

Dime-store figures were extremely successful. Millions were sold to little boys clutching their meager allowances or paper route money in their sweaty little hands. The low price made it possible for the majority of these boys to collect little armies which could fight other neighborhood generals' troops.

Because of this popularity dime-store figures are today the most collected toy soldiers in the United States. Extant old molds are now being used to reproduce some of these figures for a fraction of the price of an original and new figures are being created to supplement the armies of yore (see new old toy soldiers).

Most dime-store figures were hollow cast, made of lead alloys, but some were made of iron, rubber, and composition (see these chapters).

The companies making dime-store figures were All-Nu (page 193), American Alloy Creations (page 195), Auburn Rubber (page 310), Barclay (page 168), Grey Iron (page 214), H.B. Toys (page 195), Historical Miniatures (pages 77 and 272), Manoil (page 182), Metal Cast (page 195), Miller (page 226), Molded Products and Playwood Plastics in the United States (pages 269 and 271), Breslin (page 195), and London Toys (page 195) in Canada as well as some Japanese manufacturers (usually marked JAPAN).

As with other toy soldiers, condition is extremely important in assessing value to an individual figure.

Dime stores are often graded by quoting the amount of paint which is extant in percentage. This grading system and its equivalents in words is given below:

Mint	100 percent
Excellent +	98 to 99 percent
Excellent	94 to 97 percent
Excellent –	90 to 93 percent
Very Good +	87 to 89 percent
Very Good	84 to 86 percent
Very Good –	80 to 83 percent
Good	less than 80 percent
Average	about 50 percent

For an exhaustive treatment of dime-store figures, their history, and variations, see Richard O'Brien, *Collecting Toy Soldiers*.

BARCLAY   The Barclay Manufacturing Company, named after a street in Hoboken, New Jersey, where the factory started in the early 1920s, was the most prolific producer of dime-store figures.

By the time Barclay introduced this "new" format of 3¼ inch (82 millimeter) hollow-cast figures, it was already well established with a workforce of over one hundred boys, men, and women. Up to 1935 Barclay produced a number of smaller, 2½ inch (65 millimeter) and 2⅛ inch (54 millimeter) soldiers, cowboys, and Indians, both mounted and on foot. Some of these had movable arms and showed other signs of influence by Britains and other European makers. These were also hollow cast and unmarked. Among collectors they are known as "early Barclays." The company was blessed by an association with Woolworth's from very early on through one of Woolworth's buyers, William J. Thompson. In the early years almost the entire production by Barclay was bought by Woolworth's. Only later did other distributors such as the toy store chain Kresge's and Sears Roebuck also become of importance for Barclay. From the early years Barclay also produced cannons and cars, as well as the occasional boat, train, airplane, or even zeppelin but their dime-store figures made up by far the largest part of the production. And large it was, with Barclay employing no less than 400 workers at its peak immediately before America's entry in World War II. At this time Barclay employed forty casters, each of whom was said to be able to cast two gross (288) figures per hour. If one estimates conservatively they could produce just half that (150) for six hours a day, six days a week, the total annual production coming to ten million figures! No wonder many are left, although a large number were melted down for bullets during the war.

The first figures were designed by Frank Krupp, who in 1937 left Barclay to start All-Nu (see page 193). His position was taken by Olive Kooken, who stayed with the company until its end.

Barclay ceased production of toys after Pearl Harbor, but started up again shortly after the end of the war in 1945. In spite of modernization of the soldiers (see below), the production only reached a fraction of its prewar level and then gradually decreased during the 1950s and 60s until the company closed its doors in 1971. Original molds from Barclay are now used by Vintage Castings to produce modern versions. Vintage Castings has also made molds from original plaster models not actually put into production by Barclay.

Identification   Many Barclay's figures are not marked. They are 3¼ inch (82 millimeter) figures, however, and quite recognizable. The so-called early Barclay (2½ inches, 65 millimeters and 2⅛ inches, 54 millimeters) are far more difficult to identify as several other companies made figures of that size with rather primitive painting. The marked dime-store figures have the company name, serial number, and MADE IN THE U.S.A. as integral parts of the casting.

The Barclay dime-store figures have distinctively painted eyes; until early 1936 they consisted of curved eyelids with a dot attached underneath, giving the impression of a sideward glance. After early 1936 the dots were placed in the middle under the eyelid, giving the impression the figure is looking straight forward. From the start until about 1940 the soldiers were equipped with separate English-style steel helmets. Initially these were glued onto the head but from 1937 on were fastened with a clip attaching the helmet to the air hole on top of the figure's head. In 1940 the separate helmet was replaced by a helmet cast with the figure, so-called cast helmeted Barclays in collectors' parlance.

The figures designed by Krupp had a characteristic "short stride" with the feet close together. When Kooken took over the design, these short stride figures were largely replaced by figures with a longer, more natural stride. After World War II the early English-style helmet was replaced with the more modern version and the size somewhat reduced. In the 1950s the so-called pod-foot range was introduced, where each foot had a separate little round base. A number of the postwar pod-foot figures were painted with red uniforms and were meant to act as enemies in wargames.

Besides soldiers, Barclay made an attractive group of winter figures (skaters, sleighers, and so on), several Santa Claus figures, cowboys, Indians, and railroad passengers, the latter both in large and small scale. Most of these figures are marked in the casting.

Price   The early, smaller (2½ inch, 65 millimeter and 2⅛ inch, 54 millimeter) Barclays which are so hard to identify carry a price of about $10 each for good foot figures, about $40 for good mounted figures.

The dime-store figures vary greatly in price according to rarity and, of course, condition. Examples of prices of good, but not mint, figures are given below.

Postwar pod-foot soldiers came in khaki, later in green. Red soldiers are priced about twice or more of the value of the corresponding khaki or green figures. Most of Barclay's figures, by far, were sold singly and unboxed. Occasionally, however, they were sold in boxes or packaged in blister packs. The value of such packaged figures is about 25 percent higher than the sum of the individual figures.

Many variations in castings are known. Price for these vary according to rarity. For an exhaustive listing of such variations see Richard O'Brien, *Collecting Toy Soldiers*.

Prewar Soldiers, Cowboys, Indians, etc.

87	Officer on Horse	$20
89	Indian on horse	$20
90	Cowboy on horse	$20
701	Flag bearer, short stride	$20
701	Flag bearer, long stride	$15

Barclay: cowboy with lasso (value $40) and #89.

Barclay: #90.

701	Flag bearer, pod foot	$20
702	Machine gunner kneeling	$15
702	Machine gunner kneeling, cast helmet	$15
703	Kneeling, firing	$15
704	Marching	$10
705	At attention	$15
706	Charging, cast helmet	$20
707	Sharpshooter standing	$15
708	Officer marching	$15
709	Bugler	$15
710	Drummer	$15
711	Drum major	$20
712	Knight with shield	$10
713	Knight with flag	$15
714	Pirate with sword	$15
716	Indian chief	$10
717	Indian brave	$10

Barclay: #707, #702, #703, and #750; value $20.

Barclay: #718, #743, #760, and #746; value $15 each.

Barclay: #746,
#704, #740,
and #705;
value $10 to
$15 each.

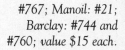
#767; Manoil: #21;
Barclay: #744 and
#760; value $15 each.

Barclay: #711,
#709, #739, and
#710; value $15
to $20 each.

Barclay: #719
and #790.

718	West Point cadet	$15
719	Sailor in white uniform	$10
720	Sailor in blue uniform	$10
720	Sailor in blue uniform, pod foot	$30
721	Naval officer	$10
722	Marine	$15
724	Ethiopian soldiers	$150
728	Machine gunner lying	$15
730	Signal man with flags	$15
731	Pigeon dispatcher	$15
732	Telephone operator	$10
733	Kneeling with shell	$10
734	Ammo carrier	$10
735	Range finder	$15
736	Sentry	$10
737	Charging with machine gun	$10
738	Throwing grenade	$10

Barclay: #733 and #952, dispatcher with dog; value $40.

Barclay: #736, #708, #704, and #765.

Barclay:
#706, #738,
#744, and
#735.

Barclay: #788,
#938, #948,
and #977.

741	Pilot	$15
744	Nurse standing with bag	$10
746	Doctor	$15
748	Soldier running	$15
749	Soldier charging with rifle	$15
754	Indian chief	$10
755	Indian kneeling with bow	$10
757	Indian standing with bow	$20
758	Cameraman kneeling	$25
759	Stretcher bearer	$15
760	Surgeon with stethoscope	$15
761	Wounded on blanket	$10
765	Bayoneting (without bayonet!)	$35
766	Clubbing with rifle	$35
771	Sitting peeling potatoes	$20
774	With anti-aircraft gun	$15
775	Wounded with crutches	$20

776	With searchlight	$25
778	Officer with gasmask	$20
779	Firing behind wall	$50
784	Parachutist landing (on feet)	$20
785	Skier in white	$20
788	Marching with slung rifle	$20
789	Anti-aircraft gunner	$20
790	Two soldiers in rubber raft	$50
791	Two man rocket launcher	$25
951	Wireless operator	$30
952	Dispatcher with dog	$50
960	Surgeon and seated soldier	$75
374	Army motorcycle with sidecar	$50

Barclay: #802, knight; value $10.

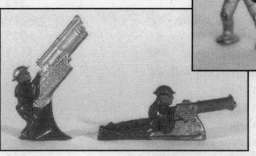

Barclay: #789 and #728.

Barclay: #791 and #790.

Postwar soldiers, cowboys, Indians, etc.

81	Two soldiers with radar (actually a listening device)	$20
82	Three soldiers with range finder	$20
83	Two soldiers with searchlight	$20
84	Two soldiers with cannon	$20
85	Two soldiers with anti-aircraft gun	$20

Pod-foot figures

901	Flag bearer	$10
903	Kneeling, firing	$10
906	Charging	$10
908	Officer	$10
909	Bugler	$10
919	Sailor in white	$10
920	Sailor in blue	$10
922	Marine	$15
928	Lying with machine gun	$10

Barclay: #83, #82, and #81.

Barclay: #909 and Midi-size figures.

937	Charging with machine gun	$15
938	Throwing hand grenade	$10
941	Pilot	$10
947	Standing firing	$10
948	Advancing	$10
950	Cowboy with pistols	$10
951	Cowboy with rifle	$10
954	Indian with tomahawk	$10
955	Indian with rifle	$10
956	Indian with spear	$10
960	Wounded on crutches	$10
961	Wounded with bandaged arm and head	$20
974	Kneeling anti-aircraft gunner	$10
977	Marching	$10
990	Bazooka	$10
991	Flamethrower	$10

Barclay: #919 and #920.

Barclay: #956, #954, #952, and #953; value $10 each.

Midi-size soldiers, similar to pod foot soldiers but smaller, are rare and the price is about $50 each. Cowboys and Indians in this size sell for $25 each.

Small Barclay railroad figures (HO scale) are priced between $5 and $10 each.

Barclay: Midi-size figures; value $50 each.

Barclay cannon; value $25.

Barclay: armored truck and truck with AA gun; value $20 each.

Barclay: truck with AA gun and truck with cannon; value $25 each.

Barclay: army truck (value $15) and truck with cannon (value $30).

Civilians

495	Man on skis	$15
496	Girl on skis	$15
497	Man on sled	$15
498	Girl on sled	$15
499	Santa Claus on sled	$30
500	Santa Claus on skis	$40
510	One horse open sleigh	$50
530	Man pulling sled with child	$50
535	Man helping girl put on skates	$75
610	Woman with dog	$10
611	Man with coat over arm	$10
612	Conductor	$10
613	Porter with broom	$10
614	Redcap with luggage	$10

Barclay: #498, #628, and #495.

Barclay: #499.

Barclay: #500.

Barclay: #510.

615	Engineer	$10
616	Boy	$10
617	Girl	$10
618	Elderly woman	$10
619	Old man	$10
620	Minister	$15
621	Newsboy	$10
622	Shoeshine	$10
623	Detective	$100
624	Burglar	$75
625	Bride	$20
626	Groom	$20
627	Girl in rocking chair	$15

Barclay: #613, #850, and #853.

Barclay: #619, #610, and #620 (two different versions).

628	Boy skater	$10
629	Girl skater	$10
635	Speed skater	$10
636	Figure skater	$10
801	Boy Scout	$30
850	Policeman	$15
851	Fireman	$20
853	Mailman	$10

MANOIL When the two brothers Maurice and Jack Manoil first started a casting business together in New York City in 1928 they called it the Man-O-Lamp Corporation in a reference to the main product, metal lamps. They also produced novelty items for chain stores. In 1934 the name was changed to Manoil Manufacturing Corporation, and the following year the company started issuing its dime-store soldiers, immediately bringing it into direct competition with the giant in the new field, Barclay.

Although never equalling Barclay in size, Manoil nevertheless held its own and at its peak in 1940 the company, which had moved to Waverly in upstate New York, employed 225 workers.

As was the case with Barclay the outbreak of World War II brought an abrupt halt to the production of the hollow-cast lead soldiers. For a period in 1944 Manoil produced some soldiers in composition (see page 269). This line was not successful as the figures much too readily disintegrated due to the inferior type of composition used.

In late 1945 production of lead soldiers resumed, but, as in the case of Barclay's, never reached the prewar level. Although a new series of soldiers in World War II uniforms was issued, the decline continued and when Jack Manoil died in 1955, the company closed its doors.

The designer of all Manoil figures was Walter Baetz. With a definite knack for caricatures, Baetz's production encompassed very lively figures indeed: the paratrooper lands on his rear end (not like Barclays' dignified GI on his feet), the motorcyclist leans over his machine in the most extreme racing fashion and the anti-aircraft gunner aims in a position only attainable by an accomplished gymnast or ballet dancer.

Most of Manoil's production was concentrated on the military, but in 1941 it issued a highly successful civilian line, the "Happy Farm" series. Like the soldiers, these civilians engage themselves in their many different endeavors with a zest unknown from other dime-store civilians. Even the scarecrow seems to be actively flailing its arms!

Identification Most early Manoil figures are marked MANOIL, MADE IN U.S.A. In 1940 the name was replaced by capital M in a circle (which had become the company's trademark). Even unmarked Manoils can be recognized from Barclays by the painting of the eyes, which consists of a line with a dot placed well underneath, a line, a dot or no eye at all. The helmet is an integral part of the figure. The very first figures issued by Manoil have a hollowed-out base, earning them the not unreasonable name Hollow Base by collectors. These figures are quite rare.

Shortly after World War II a series of extremely long-legged, spindly GIs in the characteristic World War II helmets called Skinnys was introduced. As pointed out by Henry Kurtz, these figures uncannily resembled Bill Mauldin's GI cartoon characters, Willie and Joe. Later this series was replaced by more conventionally proportioned (and smaller) GIs.

Manoil also produced a number of cars, both military and civilian, as well as cannons and other equipment. These are all marked with the company name or logo.

Walter Baetz continually tried to improve his figures, both in looks as well as in making them less breakable. For this reason a very large range of variations are found. These variations, though often very subtle, are greatly appreciated by collectors. For an exhaustive listing (and pricing) of these variations, see Richard O'Brien, *Collecting Toy Soldiers.*

Manoil tractor: value $20; armored car with cannon: value $30; and siren truck: value $50.

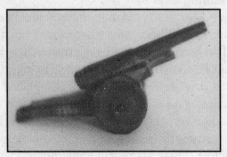

Manoil cannon; value $15.

Prices As will be seen below, Manoil prices are generally the same or slightly higher than the prices for Barclay figures. Hollow-base versions are generally two to four times the value of the flat-based versions (numbers 7 to 18). Manoil composition figures, which are quite rare because of their fragility, sell for $30 to $40 for a good, intact figure. Manoil cannons cost about $10, military vehicles $10 to $30.

The prices given below are those for good, but not mint, figures.

7	Flag bearer	$20
8	Marching man	$15
9	Marching officer	$15
10	Bugler	$15
11	Drummer	$20
12	Lying machine gunner	$20
13	West Point cadet	$15
14	Sailor	$15
15	Marine	$15
16	Naval officer	$15
17	Naval signalman with semaphores	$30
18	Cowboy with revolver	$15
18A	Cowboy surrendering	$15
20	Doctor	$20
21	Nurse	$15
22	Indian with knives	$15
23	Seated machine gunner	$20

*Manoil: #10,
#36, #31,
and #63.*

Manoil: #20, #21, #16, and #15.

24	Standing with shell	$15
25	Kneeling shooting	$15
26	Standing shooting	$15
27	Standing with tommy gun	$15
28	Kneeling with binoculars	$15
29	Wounded soldier walking	$15
30	Wounded soldier lying	$15
31	Throwing hand grenade	$15
32	Stretcher carrier	$15
33	Sitting soldier	$40
34	Pilot	$20

Manoil: #33, #82, and #76.

Manoil: #83, #58, #102, and #34.

36	Charging with rifle	$30
40	Kneeling with rifle	$40
44	Crawling with rifle	$40
46	Anti-aircraft gunner	$15
47	Searchlight	$15
48	Navy gun	$15
49	Policeman	$15
50	Soldier on bicycle	$20
52	Motorcycle	$25
57	Lying firing in grass	$20
58	Parachute jumper	$20
61	Soldier with camera	$40
63	Soldier with gasmask and flare gun	$20
65	Deep sea diver	$20
76	Rifle range target	$75
78	Kneeling with small, wheeled cannon	$25
81	Machine gunner kneeling	$25
82	AA cannon	$25
83	Trench mortar	$25
85	Pilot holding bomb	$20
86	Aviator with propeller	$65
94	Soldier running, pulling wheeled cannon	$30
102	Parachutist with machine gun	$60

Manoil: #50.

Manoil: #86; value $65.

Manoil skinnies, including #45/6.

Manoil: #45/15.

The following unnumbered figures and number 45/6 are the early postwar spindly-legged GIs (Skinnys).

	Flag bearer	$25
	Standing with submachine gun	$25
	Sitting with machine gun	$50
	Lying with machine gun	$75
	Shooting up with rifle	$50
45/6	Marching	$20
45/7	Flag bearer	$30
45/9	Standing with rifle	$20
45/11	Kneeling, shooting upwards	$25
45/12	Standing, shooting tommy gun	$25
45/13	Kneeling with bazooka	$25
45/15	General saluting on platform	$100

Figures in the 500 series are postwar thick-legged GIs in World War II helmets.

521	Flag bearer	$20
523	Soldier in poncho	$30

Manoil: #18A and #49.

Manoil: *mounted cowboy shooting; value $35.*

526	Kneeling with binoculars	$30
527	Looking straight up with binoculars	$30
530	Lying machine gunner	$25
531	Sitting machine gunner	$25
532	Kneeling firing	$25
534	Standing firing	$30

Series 41/1 to 41/41 are all of the "Happy Farm" series.

41/1	Bench	$10
41/2	Seated girl for bench	$10
41/3	Seated man for bench	$10
41/4	Man carrying sack	$15
41/6	Man sharpening scythe	$20
41/8	Farmer with scythe	$20
41/10	Sowing grain	$15
41/12	Scarecrow with hat	$20
41/13	Carrying pumpkin	$15
41/14	Eating watermelon	$60
41/19	Bricklayer	$30
41/22	Blacksmith with wheel	$20
41/23	Carpenter with door	$30
41/24	Dog	$15

Manoil: #41/22, #41/35, #41/14, and #41/24.

Manoil: #41/23 and #42/25.

Manoil: #41/33, #41/28, #41/29, and #41/31.

Manoil: #41/40, #41/8, #41/13, and #41/10.

41/25	Carpenter sawing	$20
41/28	Lady with pie	$20
41/29	Lady with child	$25
41/31	Girl watering flowers	$15
41/33	Woman with churn	$20
41/35	Woman with broom	$20
41/36	Man juggling barrel	$25
41/39	Man with water pump	$20
41/40	Boy with logs	$15
41/41	Wheat sheaves	$15

JONES   Let it first be said that there never was a company called Jones. The term (in dime-store figure enthusiasts' parlance) refers to a group of hollow-cast 3¼ inch (82 millimeter) figures made by J. Edward Jones of Chicago, IL in the late 1930s.

J. Edward Jones was an entrepreneur in the toy soldier business, leaving behind him a trail of short-lived and unsuccessful companies. The first company was Metal Arts Miniatures, started in 1929 but bankrupt in 1932. It was followed almost immediately by Miniature Products Company. This company's demise seems to have occurred in 1939. It might have been through this company that the Jones dime-store figures were issued.

However, the main production of both these prewar companies was of 54 millimeter (2⅛ inch) hollow-cast figures, some derivatives of Britains figures, others original. Their main subject was the American army from the time of the Revolution to World War I. In 1939 Jones started "Metal Miniatures" to continue this 54 millimeter production, but World War II forced it to close down in 1942.

In 1946 Moulded Miniatures Co. was started by Jones. This, the most successful of Jones's ventures, continued issuing 54 millimeter hollow-cast soldiers, especially of the American Army and Navy until 1957, not only under its own name but also under such labels as Universal, World Miniatures, Varifix, and Layout. If the purpose of all these company names was to confuse creditors or the IRS, it certainly succeeded in confusing collectors of toy soldiers!

Identification Jones, true to his penchant for obfuscation, never marked his figures other than a few early ones (saying only MADE IN CHICAGO). Identification must therefore rest on recognition of individual figures. Such recognition is clearly the realm of the super specialists and lies outside the scope of this book, although a few of the dime-store figures are briefly described below. These figures are unmarked in World War I helmets, which are an integral part of the castings.

Prices Jones' 54 millimeter (2⅛ inch) boxed sets sell for from $175 to $225. Individual 54 millimeter (2⅛ inch) figures by Jones, when recognized as such, sell for $10 to $20 each. A sampling of the most sought after of his figures, the 3¼ inch (82 millimeter) ones, is found listed below.

German kneeling firing	$150
German running with rifle	$150
German lying with machine gun	$150
U.S. officer kneeling with binoculars and rifle	$50
U.S. officer falling backwards, grabbing throat	$250
U.S. officer in overcoat with revolver, pointing	$150
Indian on rearing horse	$75
U.S. motorcyclist with machine gun	$100

Jones: a sampling of various sizes; value $15, $20, and $100.

Some U.S. soldiers are painted gray (to play the role of Germans). These are about 25 percent more expensive than the khaki variety.

ALL-NU   When Frank Krupp, the sculptor of Barclay's tin-helmeted soldiers, left that company in 1938 he started up his own, All-Nu Products Inc. The production of All-Nu dime-store figures had hardly gotten under way before Pearl Harbor and the subsequent need for other uses for lead. During the war All-Nu produced paper soldiers (see page 229) and after the war a brief, half-hearted, revival of a few civilian figures was made under the name Faben Products Inc., before Krupp turned his talents to modeling non-toy objects like lamps and clocks.

The most famous figures are the military cameraman, the "marching majorettes," and "All Girl Band," all imaginative and well-executed figures.

Identification   All-Nu figures are almost all stamped ALL-NU PRODUCTS INC. on the bottom of the base. They are, of course, strikingly similar to Barclays but with the helmet cast as an integral part of the soldier and the eyes reduced to a simple line.

Prices   All-Nu figures are much sought after, but, because of the very short production period, quite rare as reflected in the sample prices that follow.

All-Nu: girl drummer; value $100.

Military cameraman kneeling with movie camera	$2,000
Throwing grenade with slung rifle	$750
Standing firing	$75
Majorette	$500
Girl flag bearer for "All Girl Band"	$100
Girl musicians for "All Girl Band"	$100 each
Polo player on horse	$20

TOMMY TOYS   Tommy Toys was founded in 1935 but did not start producing its dime-store figures until 1937. By 1939 the company went under.

Identification   Its few soldiers are very Barclay-like in appearance (one of the designers, Olive Kooken, later joined Barclay). A unique series was the "Nursery Rhyme Figures," such as Old Mother Hubbard, Humpty Dumpty, and Jack and Jill.

All are marked TOMMY TOY under the base, some even with a characterization of the figure.

Prices   Because of their short production period and generally appealing nature, Tommy Toy figures are sought-after rarities. A few samples are given below.

Marching soldier	$125
Officer with gas mask	$250

Puss in Boots, Old Mother Witch, and Ole King Cole by Tommy Toy; value $25, $50, and $50.

Nurse with basin and towel	$150
Old Mother Hubbard	$50
Humpty Dumpty	$50
Jack and Jill (double figure)	$750

OTHER HOLLOW-CAST DIME-STORE FIGURES   Several other short-lived and generally unsuccessful companies made 3¼ inch (82 millimeter) hollow-cast lead figures both before and after World War II.

American Alloy produced outright copies of Tommy Toy soldiers which were, however, unmarked and generally of poorer quality both in regard to casting and painting. Toy Creations, which produced dime-store soldiers in the postwar period, followed the same procedure but also created a few original figures. These were not marked but many can be recognized by the copper-colored underside of the base. They are rare and quite valuable, a good figure commanding a price of about $25 to $100.

Metal Cast, originally and primarily a company producing molds for home casting, produced a few dime-store figures itself, both before and after World War II. These figures are mainly Barclay copies with minor variations. Most sell for about $10 to $30 in good condition. They are unmarked. Cowboys and Indians were also produced in the 2½ inch (63 millimeter) size, which today sells for $20 for a good figure, mounted or on foot.

H.B. Toys produced its dime-store figures from 1947 to about 1952. Although reasonably successful for a brief period, its Barclay-like figures never really took off. Their figures are actually made with a zinc rather than lead alloy. The metal is considerably harder and cannot be dug into with a knife as lead alloy can. They were unmarked. The value of H.B. Toys figures is about $15 in good condition.

During and immediately after World War II, dime-store figures were produced in Canada by Breslin Industries of Toronto, and London Toy of London, Ontario. The figures were almost all copies of Barclays and other U.S.-produced dime-store figures.

Most, if not all, are marked showing Canada as their country of manufacture. The value is the same as equivalent U.S.-produced figures.

DIME-STORE FIGURES IN OTHER MATERIALS   Grey Iron made its dime-store figures in cast iron (see page 214), Auburn in rubber (see page 310), Molded Products and Playwood Plastic in composition (see pages 269 and 271), and Beton in plastic (see page 279). Miller, sometimes considered in the dime-store group, made oversized soldiers in plaster (see page 226).

♣ NEW OLD TOY SOLDIERS

This oxymoron is used to describe model soldiers made at the present time in the style of the real old toy soldiers, especially, but far from exclusively, Britains.

Although in 1973 Britains themselves made an abortive attempt to reintroduce metal figures for the collector rather than toys, this attempt was not followed up for another ten years. Meanwhile the potential market for these figures had been recognized by Jan and Frank Scroby of England, who in 1973 introduced their new sculpted and painted figures under the trade name Blenheim. In 1974 Shamus Wade of England saw these figures and subsequently introduced his own series "Nostalgia," mainly depicting obscure and small units of the British Empire forces. These were mostly designed by Jan Scroby.

Since that time a large number of companies, mainly in England and the United States, have sprung up to satisfy an ever-enlarging market.

The only book devoted exclusively to this subject, Stuart Asquith's *The Collectors Guide to New Toy Soldiers*, from 1991, lists over one hundred such companies and lines. Many of these have only a limited range and existed only for a short period.

Almost all are solid cast in lead or lead alloy. Being relatively new (none older than twenty years), they are frequently encountered in their original boxes, making identification easy.

These boxes, usually of six figures, sell from $50 to $150, dependent on maker and subject. Single figures are harder to identify as many are not marked. Unmarked and therefore usually unidentified figures have little value, a few dollars per piece, considerably less than when they were made and part of an identified set.

Only a few manufacturers will be mentioned in this book and further information should be sought in Stuart Asquith's work or from dealers. Most toy soldier dealers also handle new old toy soldiers.

ALL THE QUEEN'S MEN   This English company started in 1981 and is still going strong. Their production covers not only Waterloo and the British Army (especially Colonial troops), but also the American Civil War.

The address is Derek Cross, The Old Cottage, Gilmorton, Lutterworth, Licestershire, LE17 5PN, United Kingdom.

ALYMER S.A.   Located in Valencia, Spain, this company dates back to 1944 although non-wargame soldiers were not produced until 1973. A large variety of soldiers have been made but the company closed its doors in 1993. The U.S. agent and very active collaborator has been K. Warren Mitchell, 1008 Forward Pass, Pataskala, OH 43062. Telephone: (614) 927-1661.

*Alymer: Arab soldier
and knight.*

BRITAINS   After the demise of Britains lead soldiers in 1966, the company produced only plastic soldiers. In 1973 they introduced a marching Scots Guard made in a rather light zinc alloy. During the next ten years only two more metal figures were introduced, a standing Life Guard and walking Yeoman Warden ("Beefeater") sold singly or in cellophane-fronted boxes of six. These figures were slightly larger than the old 54 millimeter scale. In 1983 sets of a highland officer and five men of the Gordon and Black Watch made of the same alloy were introduced. In the same year a limited edition (three thousand boxes) of eleven men and one officer of the Queen's Own Cameron Highlanders, boxed in an old-fashioned red box, was introduced. The following year Britains introduced mounted Horse Guard and Life Guard figures, later with six included in the second limited edition version. Of these seven thousand boxes were issued. Subsequently Britains has issued a number of different sets of figures from different regiments, both in unlimited and limited editions.

Britains:
Bahamas Police
Band from the
limited edition
series.

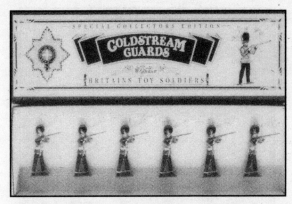

Britains: copies of earlier lead figures—infantry.

Britains: copies of earlier lead figures—cavalry.

In 1989 boxed sets (six infantry and four cavalry) of copies of lead figures were introduced. These editions are not limited in number.

The address is Britains Ltd., Chelsea Street, New Basfard, Nottingham, NG7 7HR, United Kingdom.

GUARD CORPS, LTD. This is the latest of the new old toy soldier companies. It has started producing some very well designed Civil War figures as well as German Imperial troops. The soldiers are marked GC.

The address is Ed Lober, 71-35 71st Place, Glendale, NY 11385.

Guard Corps boxed set of Union troops.

Guard Corps boxed set of Union troops.

*Guard Corps Union
standard bearer.*

Guard Corps Prussian grenadiers.

WILLIAM HOCKER   In my personal estimation, this is the producer of the most attractive new old toy soldiers. Started in 1983, the company has concentrated on making sets that "Britains should have made, but didn't." Many of the nineteenth century wars of the British Empire are treated very elegantly and the figures go well with Britains displays.

The address is William Hocker, 1605 Arch Street, Berkeley, CA 94709.

William Hocker Boer War British balloon corps.

William Hocker Boer War British naval gun team.

William Hocker box.

IMPERIAL PRODUCTION   This New Zealand company was formed in 1982 and has established itself to be among the best, if not the best, producer of new old toy soldiers. The subjects covered are mainly scenes from the history of the British Empire troops, but the American Civil War has recently been added to the repertoire. The figures are marked IPNZ.

The address is David Cowe, P.O. Box 94, Greytown, New Zealand.

Imperial Production mounted officers.

KING AND COUNTRY   This Hong Kong-based company was founded in 1985 by Andy Neilson and Laura McAllister. The production covers mainly the late Victorian era. The soldiers are unmarked.

The address is Andy Neilson and Laura McAllister, Ground Floor, 31 Wyndham Street, Central, Hong Kong.

MARLBOROUGH MILITARY MODELS, LTD.   Like Phoenix, this company rose from the ruins of Blenheim. Owned by Frank and Jan Scroby, it has produced a large number of civilian figures from the Victorian era, both English and from the Empire of India. The range also includes the spectacular Delhi Dunbar of 1902 including exciting items like elephants and camels. The figures are marked MMM.

The address is Frank and Jan Scroby, The Duchy, Pontycymmer, Bridgend, CF32 8DU, United Kingdom.

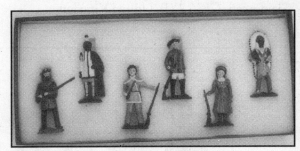

Marlborough set of early American heroes.

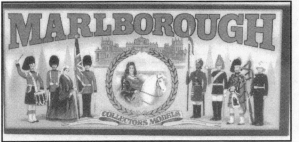

Marlborough box.

NOSTALGIA MODELS   Started by Shamus Wade in 1974, this company closed ten years later. In the meantime it produced a large series of soldiers comprising depictions of small and exotic units of the British Empire troops and police forces.

PARADE MARCH TOY SOLDIERS   This company, owned by Neal Crowley, produces 10-centimeter (4 inch) Elastolin-like figures in resin. The address is 535 Barrington Road, Grosse Pointe Park, MI 48230.

TROPHY MINIATURES WALES, LTD.   This company, which started in 1972 but did not start "mass production" until 1978, was formed by Len Taylor. Its production includes such exotic items as a Nile River Boat, Arab dhow, steam tractor, and vintage cars to display personality figures. The figures are unmarked.

The address is Len Taylor, Units 10-22, Barry Workshops, Sully Moors Road, Sully, Penarth, South Glamorgan, CF26 2XB, United Kingdom.

Parade March Toy Soldiers.

II

𝔄luminum

❖ Aluminum was used in the manufacture of toy soldiers in France even before World War II. Quiralu and Frenchal were prominent, but even Mignot experimented with aluminum in the 1930s under the name Mignalu. Garrett (*The World Encyclopedia of Model Soldiers*, London, 1981) mentions that Beton produced aluminum figures, but I have been unable to find any confirmation of that statement.

After World War II, Wend-Al of England produced a number of aluminum figures in cooperation with Quiralu. I have also found examples of Japanese aluminum figures, all copies of Elastolin 10-centimeter (4-inch) or 7-centimeter (2¹³⁄₁₆-inch) figures. The time of their manufacture is not known, but is likely prior to World War II. Other companies might also have used

Frenchal:
papal guard;
value $10.

Frenchal: West
Point cadet;
value $8.

Various French soldiers by Quarilu; value $10 each.

Quarilu: French Chasseur d'Alpin; value $15 each.

aluminum, the main advantages of which are its light weight and durability. Its drawbacks are a high melting point and difficulties in obtaining truly detailed features in the castings.

In Denmark after World War II a number of aluminum soldiers were produced. The most successful company was Krolyn, started in the late 1940s and in production until 1958.

The range of figures by Krolyn fall into three main categories: cowboys and Indians, Vikings, and the Robin Hood Series. The latter included knights, Ivanhoe, and at times even some of the Viking figures.

The cowboys and Indians were quite unoriginal copies. Indian horses by both Elastolin and Lineol were used, but there are indications that only Lineol's horses were used for this series, the Elastolin horses being reserved for the mounted knights. Cowboys used the Lineol Indian mount. One of these, the Elastolin figure of a mounted cowboy shooting his pistol, was painted as a 7th Cavalry man. The only original figure I know of in this series is a cowboy shooting his rifle from behind a rock, an obvious modification of Elastolin's Indian in the same position.

Far more original are the Vikings by Krolyn. Their positions are original and imaginative, well sculpted and painted. My

Quarilu: French machinegunner; value $15.

Quarilu: French Chasseur d'Alpin on skis; value $20.

Wend-al box.

Wend-al: Life Guard drummer; value $20.

Wend-al: Life Guard bugler; value $15.

Wend-al: Horse Guard standard bearer; value $15.

favorite is the berserk holding his enormous club above his head. He is reminiscent of one of the "wild men" flanking the Danish Royal Arms. Many Vikings are marked RØDE ORM[1] on the base (together with a number used as reference by Krolyn).

The name refers to a novel popular in Denmark at the time titled RØDE ORM and written by the Swede Frans Gunnar Bengtson (1894–1954).

Krolyn: viking;
value $30.

Krolyn: vikings; value $25 each.

---

[1] Meaning "red snake." The snake was revered by the Vikings.

The same strategy was used for the so-called Robin Hood series, which actually covered a wide range of similarly well-known characters, fictional or real, from the romantic era of knights. Besides Robin Hood himself, Little John, Will Stutely, Guy of Gisborne, Ivanhoe, Friar Tuck, and Richard the Lionhearted are examples. The figures of this group are mainly original or greatly modified pre-existing ones. An example of the latter is the figure of Robin Hood with his bow, modified from the Lineol figure of the Indian in the same position. The horses used for the mounted knights were all copies of the Elastolin jumping Indian horse.

Wend-al: Salvation Army band; $125.

Krolyn: Robin Hood and Friar Tuck; value $25 each.

Wend-al: snake charmer (the snake is Britains'); value $15.

Krolyn figures occasionally pop up at American toy soldiers shows and in antique stores. They are generally wildly priced (I have seen a paintless Indian for $60!). In Denmark, where these figures are far more common, the price for poor quality figures (meaning with little paint left—the figures themselves are virtually indestructible) is usually $5 to $10. Figures with preserved paint may cost anywhere from $15 to $50. The most desirable figures are the ones of the Viking and Robin Hood series (including knights), unusual subjects for toy soldier manufacturers.

Other aluminum soldiers sell in the $5 to $15 range.

Krolyn: knight;
value $40.

Krolyn: knight;
value $40.

Krolyn: Hamlet;
value $30.

Krolyn: Royal
Canadian Mounted
Police; value $25.

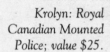

Brass

❧ Occasional soldiers made of unpainted brass appear. In style they are almost abstract—Art Deco-like in appearance and generally in the 2 inch (50 millimeter) to 2½ inch (62 millimeter) size.

These figures appear to have been made in China in the interwar period.

Few people collect these figures and they would sell for $3 to $8 each.

Unpainted Chinese soldiers made of brass, presumably made in China in the 1930s; value $5 each.

Cast Iron

✤ The Americans have traditionally made cast-iron toys, vehicles in particular. It is therefore little wonder that the only toy soldier makers who have been brave enough to use this hard metal, not very suitable for casting details and so easily oxidized, have been American. Despite their drawbacks, these solid-cast heavy soldiers enjoyed a not insignificant degree of success before they, like lead soldiers, were replaced by plastic.

✤ GREY IRON

The first dime-store figures (see page 167) were introduced by Grey Iron in 1933. The company had started issuing nickel-plated 1½ inch (40 millimeter) Grey Klip soldiers in 1917, a production which continued into 1942. In the latter part of this period these soldiers were painted khaki. The term Grey Klip refers to the fact that the base is made to slide into a metal clip. These soldiers were sold in sets, cleverly mounted on a piece of cardboard. The dime-store figures were introduced in 1933 and were produced until the beginning of World War II. Most Grey Iron's dime-store figures were military, but in 1940 they introduced the "American Family" series, including both homelife and the family on the farm (with appropriate animals), on the beach, and on the ranch. These civilians were smaller (2¼ inches, 58 millimeters) than the regular 3¼ inch (82 millimeter) Grey Iron dime-store figures.

Grey Iron was founded about 1840 in Mount Joy, PA. Over the years the company has produced a plethora of cast-iron objects, from door hinges to stovepipes. The company still exists, although now under the name Donsco Inc., acquired when it was combined with several other casting companies in 1974. The Grey Iron division, now called John Wright, is located in Wrightsville, PA.

Since the end of World War II Grey Iron (and John Wright) has intermittently produced many of its prewar dime-store figures. Unpainted castings are available at the factory. At a neighboring factory, Wilton, some cast-iron figures were also produced in small numbers. They are virtually indistinguishable from Grey Iron, who copied them.

Identification   Grey Iron is the only company producing cast-iron dime-store-size figures in quantity. Therefore, the material itself identifies the piece. The figures are unmarked. They are rather stiff looking, with relatively little detail. The "American Family" series contains some quite unusual pieces such as a man in a bathing suit lying on his back with a newspaper over his head, a girl skipping rope, and a dude on a bucking bronco.

Prices   Grey Klip soldiers (1½ inches, 38 millimeters) sell for about $5 each, whereas sets in mint condition on the original cardboard may sell for about $100.

Grey Iron dime-store figures vary considerably in price (see sample list below). The earlier models, slightly smaller than the later, are generally the most valuable. A series of Foreign Legion figures (actually doughboys in blue uniforms) cost about the same as their equivalent in khaki. The "American Family" figures (which incidentally fit O scale trains) are generally of lesser value (see below).

The list below represents prices of Grey Iron dime-store figures in good, but not mint, condition.

1	Colonial soldier	$20
1MA	Colonial mounted officer	$40
2	West Point Cadet	$15
2A	West Point Cadet officer	$20
3	U.S. infantry man (in Montana hat)	$15
3A	U.S. infantry officer (two versions)	$15
4A	Doughboy (in helmet) kneeling with binoculars	$20
4/1	Doughboy signaling with flags	$20
4/3	Doughboy lying with rangefinder	$75
4/5	Doughboy kneeling shooting	$20
4/6	Doughboy attacking with bayonet	$15
5	U.S. infantry advancing with rifle	$15
6	Doughboy marching	$10
6A	Doughboy officer	$10
6/2	Doughboy standing guard in coat	$20

Grey Iron: #2A, #14, #14A, and #3A.

6/3	Doughboy crawling with grenade	$25
8M	U.S. Cavalry man	$30
9	U.S. Marine	$15
10	Royal Canadian Mounted Police standing	$25
10M	Royal Canadian Mounted Police mounted	$35
11	Indian with tomahawk	$15
11/1	Indian scouting	$20
11/2	Indian attacking with tomahawk over head	$75
11M	Mounted Indian	$25
12	Cowboy standing	$10
12/1	Hold-up man	$15
12/2	Cowboy with lasso	$30
12/2	Cowboy surrendering	$75
12M	Mounted cowboy	$35

Grey Iron: #21, #6, #5, and #8.

Grey Iron: #11/1, #12/1, #12, and #12/2.

Grey Iron cannon: value $10; #13, and #6/3.

Grey Iron: #18/1 and #20.

Grey Iron: #D27.

13	Doughboy kneeling with machine gun	$15
13/1	Doughboy lying with machine gun	$15
14	U.S. sailor	$15
14A	U.S. naval officer	$15
14/1W	U.S. sailor signaling	$25
15/1	Boy Scout saluting	$15
15/2	Boy Scout walking	$15
16/1	Pirate boy	$20
16/2	Pirate with two pistols	$20
16/3	Pirate with knife	$20
16/5	Pirate with cutlass	$15
17/1	Legion drum major	$20
17/2	Legion bugler	$15
17/3	Legion drummer	$15

17/4	Legion flag bearer	$15
18/1	Ethiopian tribesman advancing	$50
18/2	Ethiopian with sword	$50
18/3	Ethiopian marching	$30
19	Knight in armor	$15
20	Doctor	$25
21	Doughboy stretcher bearer	$30
22	Wounded on stretcher	$25
22/2	Wounded on crutches	$30
23	Nurse in short cape	$20
6F	Foreign Legion marching (blue uniform, same casting as **6**)	$25
6AF	Foreign Legion Officer (same casting as **6A**)	$25
D27	Doughboy with wounded comrade	$200
6/3F	Foreign Legion crawling with grenade (same casting as **6/3**)	$30
8/F	Foreign Legion mounted	$45

American Family

T1	Man with suitcase	$10
T2	Woman	$10
T3	Boy	$10
T4	Girl	$10
T6	Engineer	$10
T8	Policeman	$10
T10	Newsboy	$10
T11	Preacher	$10
F1	Farmer	$5
F4	Man digging	$5
F5	Horse	$5
F6	Cow	$5
F12	Dog	$5
H1	Man with watering can	$10
H3	Boy with kite	$15

H4	Girl skipping rope	$15
H5	Old man sitting	$5
H6	Old woman sitting	$5
H11	Milkman	$10
B1	Man lying in bathing suit	$25
B2	Woman sitting in bathing suit	$25
B6	Boy with life preserver	$25
B7	Girl with sandpail	$25
R2	Dude for bucking bronco	$15
R8	Bucking bronco	$15

♣ ARCADE

The only other company to venture into toy soldier production in cast iron was the company Arcade, best known for its cast-iron vehicles. In the late 1930s, Arcade produced a few approximately 1½ inch (40 millimeter) soldiers so Art Deco in style they are unmistakable. They were made in nickel-plated versions and sell for about $5 each.

V

Lithographed Tin

✤ Two types of lithographed tin soldiers have been made, one where the tin is flat and one where the tin is molded to make a fully (or halfway) round figure.

FLAT LITHOGRAPH TIN

The flat type was produced by Marx in the 1930s. They were made as part of sets which would also include a popgun or cannon. The tin was bent at the bottom at a slightly acute angle so the soldier could stand vertically but easily toppled with a well-aimed shot.

The soldiers are 3 inches to 4½ inches (75 millimeters to 115 millimeters) tall. Most are U.S. soldiers in action in the uniforms of the time, but other countries, cowboys and Indians, tanks, sailors, and warships are also represented.

Marx: *flat lithographed tin soldiers; value $5 each.*

Marx: *flat lithographed tin soldiers; value $5 each.*

Marx: *flat lithographed tin soldiers; value $5 each.*

Identification   The figures are marked with Marx's trademark.

Prices   Most single soldiers sell for $5 to $6. Some rarer ones, like the ski trooper on patrol, may cost twice that. Complete boxed sets with gun or cannon cost $150 to $200.

MOLDED LITHOGRAPH TIN

Although this process, where the tin is formed over a matrix (one for each side) and then joined, is mainly used for wheeled vehicles, it has also been employed in making soldiers. The process has been known since about 1840 but most are from the first quarter of the twentieth century. Earlier figures were hand painted. Generally these soldiers are 60 to 80 millimeters (2⅜ inches to 3⅛ inches) in height, not fully rounded in width [usually less than 10 millimeters (⅜ inch)], and stand on a separate formed tin plate. Such soldiers are known to have been made by EBO in Germany, M. Laas of Paris, and Paya of Spain. The soldiers are in uniforms of the turn of the century, most are marching, but some in action. Reproduction from old dyes dating back to 1933 have been issued by Paya in recent years. Both old and new reissues cost $25 to $50 if in very good condition. Boxed sets are rare and the price is not established.

EBO: molded lithographed tin soldiers; value $30 each.

Large, Japanese-made molded lithographed tin soldier with mechanical machine-gun; value $100.

Paya: reissue of Spanish cavalry; value $30.

VI

Plaster

✤ This unpromising material has been used in various countries and at various times for making toy soldiers. The fragility and lack of water resistance of plaster has however doomed large-scale success as a toy for other than the most fastidious of children.

German, Italian, Japanese, and French companies experimented with plaster in the earlier part of the century but only few and generally very poor figures have survived to this day. None can be considered collector's items but merely curiosities at best. Most common are Italian creche figures which sell for a few dollars each.

In the United States two companies reached a certain limited success in using plaster.

✤ MILLER

J.H. Miller Company was started in 1938 in Chicago. Until the Korean War production was limited to creche figures. During the Korean War, as the supply of lead for toy soldiers decreased drastically in the United States, Miller had its chance to replace some of the dime-store figures in the five- and ten-cent stores. These soldiers are therefore considered dime-store figures, although they were considerably larger (5 inches or 128 millimeters tall) than the conventional type.

By 1959 plastic soldiers forced the company into bankruptcy. Although well designed, the fragility of the figures was a definite limitation for their popularity as well as survival rate.

Identification Slik-Toy (see page 228) is the only other soldier of this size (5 inches) produced in plaster but these are clearly marked SLIK TOY. Millers are marked MILLER 1950 or MILLER 1951. They are in Korean War uniforms.

Prices The weapons for Miller figures were, for the most part, cast separately and are therefore often missing. Their limited popularity when made also has contributed to a limited attractiveness as a collectible. The prices listed below refer to good but not perfect figures. Miller's nativity figures (which are unmarked, but typically have a paper tab protruding from the foot) sell for $5 to $10 each.

Miller: MacArthur and nurse.

General MacArthur	$35
Nurse	$25
Marching with rifle	$15
Marching with flag	$15
Advancing with rifle	$15
Planting flag (à la Iwo Jima)	$15
Standing with binoculars	$15
Throwing hand grenade	$15
Kneeling with flamethrower	$50
Kneeling with bazooka	$15
Kneeling with radio	$15
Lying with rifle	$15
Lying with bazooka	$15
In foxhole with rifle	$15

Miller: GI advancing with rifle and throwing handgrenade.

Miller: GI lying with rifle.

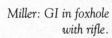

Miller: GI in foxhole with rifle.

✤ SLIK-TOY

During the Korean War Lansing Co. Inc. of Lansing, Iowa, started producing 5 inch (127 millimeter) tall GIs in plaster. Production lasted until 1956. These figures, reminiscent of Miller's, were produced in smaller numbers and therefore rarer.

Identification   Quite similar to Miller's but with the name SLIK TOY carved into the base, these figures are readily recognizable. Weapons were cast as an integral part of the figures. Only half a dozen different figures were made.

Prices   Because of their scarcity, Slik-Toy soldiers are more expensive than the equivalent Miller figures. Depending on condition, price is $25 to $50 each.

✤ GARDEL INDUSTRIES

Two series of plaster soldiers were made by this company, both the traditional 3¼ inch (82 millimeter) size, "United Nations Forces," a group of about half a dozen different soldiers, and a vastly undersized tank were unpainted. The "Empire Forces," four soldiers and two undersized tanks, were painted green. They included an oversized standing paratrooper. All were made during the war years only. They were sold in boxed sets, United Nations Forces with watercolors and brush for painting.

Identification   The combination of material and size identifies these quite rare and unattractive figures.

Prices   Although rare, the unattractiveness of the figures limits their value. $10 to $20 is probably an average price.

VII

Paper and Cardboard Soldiers

❧ The very first mass-produced toy soldiers were made of paper in the middle of the eighteenth century. First issued in France, the idea caught on rapidly in Europe during the Napoleonic era. Some were issued in black and white only, others hand colored. The invention of color lithography in the middle of the nineteenth century, and its wide distribution in the latter half of that century, led to a very large production of paper and cardboard soldiers at that time. Most often paper soldiers were issued in sheets to be cut out and pasted on cardboard by the child himself. Cardboard of various types was also used and in some cases, especially in the United States, the soldiers were cut out and placed on wooden stands by the printer or distributor. Although the last part of the nineteenth century and earliest part of the twentieth saw the greatest popularity of paper and cardboard soldiers, they experienced a renaissance during World War II due to the metal shortage. Paper soldiers have been made in almost all European countries and some are still being made today.

The subjects covered by paper and cardboard soldiers are myriad. The printing was cheap and variations literally infinite. During World War II punch-out cardboard soldiers, airplanes, warships, and vehicles became especially popular, although they did not survive the return of metal and subsequently plastic.

Collectors of paper and cardboard soldiers much prefer them in their original state, which for paper soldiers means they have not been cut out and mounted and for the punch-out cardboard soldiers that they have not been punched out and assembled. A few American publishers stand out in the production of paper and cardboard soldiers.

The size of cardboard and paper soldiers varies greatly. The bicycling figure is by an unknown European publisher, the others by unknown U.S. publisher(s); value $2, $15, and $3.

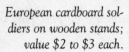

European cardboard soldiers on wooden stands; value $2 to $3 each.

European cardboard soldiers on wooden stands; value $2 each.

European hand-painted cardboard soldier; value $5.

Cavalry by an unknown European publisher; value $1 to $2 each.

✤ McLoughlin

This New York publisher of children's books and board games was very prolific in its production of paper and cardboard soldiers both in sheets and mounted on cardboard and sold in boxes. The size of the soldiers varies from 4 inches to 6 inches (100 millimeters to 152 millimeters).

Identification Both boxes and sheets are marked with the publisher's name, but individual soldiers are not. Cut-out soldiers are therefore quite difficult to identify. In the earliest part of the twentieth century a particularly rich type of chromolithography was used, the so-called glossy series.

Prices Boxed sets vary in price from $100 to $300, dependent on the number of figures included. Individual sheets

McLoughlin: *zouaves;*
value $1 each.

McLoughlin:
zouaves; value
$1 each.

McLoughlin: U.S. Navy; value $2 to $3 each.

or strips cost $15 to $50. Individual cut-out soldiers vary in price from less than a dollar to $5, dependent on size and condition. Factory-mounted soldiers are more valuable than the home-mounted variety.

✤ MILTON BRADLEY

An early producer of board games, Milton Bradley of Springfield, MA, started producing mounted cardboard soldiers at the turn of the century. These were usually mounted on wooden bases and of the 6 inch (152 millimeter) scale, selling in boxes of various sizes. Although generally not as attractive as McLoughlin's glossy soldiers, they are nevertheless of high quality.

Identification Boxes are labeled with the company's name.

Prices Boxed sets sell for $50 to $100. Individual mounted figures sell for $2 to $3.

Other U.S. companies that were active at the turn of the century were Parker Bros. of Salem, MA, Litho Novelty Co. of New York, Samuel Gabriel Sons & Co. of New York, and Clark and Sowden of New York, all producing high quality paper and cardboard soldiers.

Several well established toy companies switched their production from metal to cardboard with the entrance of the

United States in World War II. Marx issued a set of soldiers in cardboard with a cardboard gun (to shoot rubber bands) as a substitute for its lithographed tin soldier sets. Wyandotte made vehicles and airplanes for assembly as well as cardboard soldiers. Built-Rite made similar sets with forts for assembly and All-Nu, which had produced a few dime-store figures prewar, made decal sheets for mounting on cardboard.

There are several collectors of these war toys but generally the prices are quite low.

McLoughlin: U.S. Navy; value $1 to $2 each.

McLoughlin: U.S. sailors; value $8 for the set.

McLoughlin:
U.S. artillery;
value $5.

McLoughlin: U.S.
soldiers; value $3
each.

VIII

Composition Figures

❧ The early forerunners of what we today call composition figures were made of material of very limited durability, in the eighteenth century of rye flour and later papier-mâché and plaster mixtures. Around 1850 a material consisting of a mixture of sawdust and glue was introduced in Germany for the production of dolls' heads and limbs. However, because of the very high water content, problems often resulted from irregular shrinkage. Later in the nineteenth century the Viennese toy manufacturer Emil Pfeifer added kaolin (and possibly other compounds such as casein and dextrin) to the mixture and applied relatively high pressure to heated molds. This made the production of durable composition figures possible. The composition mixture is usually applied over a steel wire armature, pressed into the multipart mold, and baked while still in the mold. When removed from the mold the figure is trimmed and then painted.

Emil Pfeifer, whose main products were dolls' heads, started experimenting with the manufacturing of toy soldiers made in this very cheap and readily available material in the 1890s. The experiments were both technically and commercially successful and his process was soon emulated by other toy manufacturers, especially in Germany.

Emil Pfeifer's figures are quite large (about 11.5 centimeters—4½ inches) often with detailed molding and even more detailed painting. Special attention was given to facial features, probably because of Pfeifer's earlier experience painting dolls' heads. The Pfeifer figures thought to be the earliest are proportioned more like children than the adults they are meant to depict. Their very large heads give them a doll-like appearance. This error in proportion was soon corrected and more adult, often very lifelike, figures made. Initially Pfeifer figures were made without a stand, relying on their rather large feet for support. As anyone trying to line these soldiers up will know, this was clearly unsatisfactory. A shake of the table sends them tumbling over, often with a domino effect. Experiments with a base made of pressed cardboard were not very successful, since the

Pfeifer figures: two very early, large-headed, doll-like figures are at the far left; value $50 each.

base tended to detach from the protruding armature of the feet to which it was attached. The solution was found in making the base an integral part of the composition figure. Not only did this simple solution, albeit somewhat clumsy in appearance, allow better stability but it also offered the opportunity to make figures in far more animated positions than simply marching or standing still. Soldiers, cowboys, and Indians, as well as their horses, could be shown engaging in virtually any martial or peaceful activity. In the early 1900s Pfeifer changed the name of his composition figure department to Tipple-Topple—perhaps an allusion to their early instability. In 1925 Pfeifer sold this part of his business to O. & M. Hausser (Elastolin) and the production of these two companies was partially integrated by an exchange of molds and ideas for new products.

O. & M. Hausser, originally Miller & Freyer, started production of composition figures in 1904. In 1912 they decided, as a sales ploy, to call their composition Elastolin. This turned out

Pfeifer: U.S. Spanish War soldiers. Value: mounted figures $150; foot $40 each.

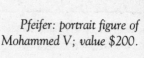

Pfeifer: portrait figure of Mohammed V; value $200.

Pfeifer: early Austrian soldier on cardboard stand; value $75.

Pfeifer: Indians on composition stands. Value: mounted figures $100; foot $40.

to be an ingenious idea. The name caught on and became almost synonymous with all composition figures, but gave Hausser the advantage of being the maker of "real" Elastolin figures (much as aspirin, Bayer's trade name for acetylsalicylic acid, became the common name for a medication). In 1903, Oskar Wiederholz had also started experimenting with composition. After some early frustrations his experiments were crowned with success. In 1906 he started using the trade name Lineol, a name which became almost as famous as Elastolin. Elastolin and Lineol were in continuous competition until the end of the composition era in the 1950s.

Initially both Lineol and Elastolin made figures of a bewildering number of sizes. Lineol for instance made early figures of 18 (7⅟₁₆ inches), 14 (5½ inches), 10 (4 inches), and 8.5 centimeters (3⅜ inches) in size. These were originally standing on a wooden base to which they were attached by an elongation

Tipple-Topple: portrait figure of Robinson Crusoe's helper, Friday; value $200.

Elastolin: 10-centimeter, mounted U.S. officer; value $100.

Elastolin: 7-centimeter and 10-centimeter Royal Danish Guard; value $40 each.

of the steel wire armature. Later the base became an integral part of the composition figure, not only simplifying the production but also allowing the trade name to be embossed underneath the base. The company soon settled on a standard size of 8.5 centimeters (3⅜ inches) [sometimes called 9 centimeters (3⁹⁄₁₆ inches)], a size in which they made figures until the very early 1930s. Elastolin developed along the same lines, but settled for a standard size of 10.5 centimeters (4⅛ inches) (often called 10 centimeters) for the standing foot figure. Most of the competing companies emerging in the teens and twenties of this century followed Elastolin's lead regarding size. The majority of these companies, formed in Germany and neighboring countries, had a relatively short life span and most of their figures are of definitely inferior quality. More than one hundred companies producing composition figures are known, but many only by advertisements in trade magazines or from single figures.

Although in their early years both Lineol and Elastolin made figures on square as well as oval bases, a convention by which Elastolin made theirs on oval bases and Lineol utilized square bases, was soon established, facilitating the recognition of products from these main producers of composition figures.

During the 1920s both Elastolin and Lineol introduced a smaller and cheaper size for their figures varying between 5 (2 inches) and 6.5 centimeters (2⁹⁄₁₆ inches). For a time these figures were produced simultaneously with the larger size figures, but in the early 1930s the two sizes were merged by both companies into what is today considered the standard 7 (2¾ inch) or 7.5 centimeter (2¹⁵⁄₁₆ inch) size (½₅th scale). Shortly before World War II they both produced a small number of 4 centimeter (1⅝ inch) soldiers, but the war inhibited any development of this size.

Although most composition figures are of full three-dimensional configuration, semi-round figures, popular on the continent and in the United States in lead, especially for home casting, were also produced in the early years of composition. Lineol briefly experimented with these figures. Best known though are the so-called Manz figures by an unknown manufacturer. A fair number of these, in a large variety of uniforms representing many different nations, do pop up both in the United States and in Europe. They had the advantage of not needing steel wire armatures, therefore being easier and cheaper to produce, but this was not enough to enable them to compete with more realistic fully rounded figures.

Composition figures never really caught on in England where hollow-cast figures so totally dominated the market, but

Left: Lineol 7-centimeter, 6.5-centimeter, 6-centimeter, and 40-millimeter size German soldiers. Right: Elastolin 7-centimeter, 6.5-centimeter, 6-centimeter, and 40 millimeter German soldiers.

TAG figure. TAG is one of only two English manufacturers to make composition soldiers; the other is Bell. Value $15.

on the continent these figures gained great popularity. In the United States they never became more popular than the metal Britains. Among soldier collectors composition figures are now gaining in popularity after having been regarded with disdain for many years.

♣ LINEOL

Lineol soldiers were produced from 1906 to the early 1940s. After World War II production of animals, cowboys, and Indians was continued in Dresden, East Germany, where the factory moved from Brandenburg. During this postwar period a small number of undistinguished East German soldiers were also produced. This factory closed in 1958. Gert Duscha started the reproduction of Lineol figures in resin in 1988 and others have since made them in metal. The latter are easily distinguished by the material used. The plastic figures, which also include new figures and positions, are best told by the lack of airholes on the bottom (invariably present in composition figures as the easy escape of steam was essential for the manufacturing process).

Identification   Lineol figures are generally all clearly marked on their square foot plate with the name LINEOL and GERMANY. Figures without foot plates (kneeling, mounted detachable riders) are not marked. Most animals are marked by the name or just GERMANY in the casting on the belly or other suitable places. Some animals are unmarked. Cowboys, Indians, and civilians do not normally have square bases. Unmarked square bases, unless on very early figures, indicate the figures are not made by Lineol.

Figures made before about 1932 are mainly of 9 centimeter (3⁹⁄₁₆ inch) or 6.5 centimeter (2⁹⁄₁₆ inch) sizes. These figures are generally quite stiff looking and have little interest for collectors. Figures made after about 1932 are often 7 centimeters (2¾ inches) in size and are the most widely collected. The 4 cen-

Lineol: U.S. soldiers in 6.5 centimeter, 7 centimeter, and 9 centimeter size. Value $5, $15, and $10 respectively.

*Lineol: 9-centimeter
U.S. medic with dog;
value $50.*

timeter (1⅝ inch) figures, exquisite as they may be, are few and far between and seldom collected.

Prices The value of 9 centimeter (3⁹⁄₁₆ inch) figures is generally not high. Marching men in close to mint condition sell for about $10 apiece, mounted for $20 to $25. Action figures are more valuable but less so than Elastolin figures. Horsedrawn tin vehicles (usually ambulances or cannons) fetch $2–$350 retail. The 6.5 centimeter (2⁹⁄₁₆ inch) figures are not of any great value either, marching figures fetching $5–$10, mounted $15–$20, and horsedrawn vehicles $75–$150. Unusual positions might of course be of greater value, but the collectors of these are few.

The most valuable Lineol figures are those of the 7 centimeter (2¾ inch) size, personality figures in particular. The retail prices of a selection of these figures in good condition (minor paint loss, hairline cracks only) are given below. Lineol made figures of many different nations. German, English, and American-uniformed figures tend to be the most valuable.

5/1	Hitler standing saluting	$150
5/2	Hitler walking saluting	$200
5/3	Hindenburg standing in spiked helmet	$100
5/4	Hindenburg standing in coat	$125
5/5	Ludendorff standing in cape	$125
5/6	von Mackensen in Hussar's uniform	$125

#5/1 and #5/2.

#5/3, #5/4, #5/5, #5/6, and #5/7/1.

5/7/1	von Blomberg saluting with baton	$125
5/8	General in coat	$40
5/10	Officer with right arm akimbo	$35
5/12	Officer with map	$35
5/13	Officer with binoculars	$35
5/24	Mounted officer attacking	$75
5/30	Marching flag bearer with flag	$75
5/37T	Marching man with backpack	$15
5/37O	Marching officer	$25

#5/8, #5/10, and #5/12.

#5/24.

#5/30, #5/37T,
and #5/52.

5/41 to		
5/56	Marching musicians	$35 each
5/62	Officer attacking with sword	$35
5/64	Man advancing with rifle	$30
5/65	Man attacking with rifle	$30
5/66	Man standing shooting	$25
5/67	Man kneeling shooting	$25
5/68	Man lying shooting	$25
5/72	Man with flamethrower	$60
5/77	Man clubbing	$30
5/76/1	Man lying with hand grenade	$35
5/76/2	Man kneeling with hand grenade	$40
5/79/2	Man standing throwing hand grenade	$30
5/87	Man falling backwards	$35
5/110/2	Man standing with shell	$40
5/149	Man lying with machine gun	$40
5/150	Man lying with ammo belt	$35
5/158	Man lying with ammo boxes	$35

#5/62, #5/64,
and #5/65.

#5/72.

#5/77,
#5/76/2,
and
#5/79/2.

#5/87 and
#5/163.

5/163	Man advancing with light machine gun	$50
5/118	Man with semaphores	$40
5/177	Man with searchlight	$45
5/174	Man walking a dog	$60
5/100	Man with radio with antenna	$50
5/171	Man with carrier pigeons	$60
5/233	Man kneeling with telephone	$35
5/211	Man with motorcycle	$175
5/180	Doctor	$35
5/182	Nurse with bandage	$35
5/204	Stretcher-bearer team with stretcher	$85

#5/118 and #5/177.

#5/174 and #5/100.

#5/171 and #5/233.

5/222	Man sitting eating	$35
5/225	Man washing face	$40
5/251	Mounted soldier	$85
5/251O	Mounted officer	$85
8/37	Sailor marching	$60
7/2	Goering in Air Force uniform and coat	$150

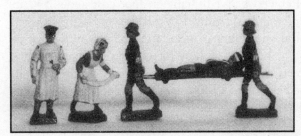

#5/180, #5/182, *and* #5/204.

#5/211.

#5/225 *and* #5/251.

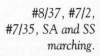

#8/37, #7/2, #7/35, SA *and* SS *marching.*

7/37	Luftwaffe man marching with	
	slung rifle	$40
	SA marching	$35
	SS marching	$40
	SA musicians	$100 each
	SS musicians	$125 each

Lineol cowboys and Indians on foot cost about $40 in good condition, mounted figures $75–$100.

Lineol animals vary in price from $5 for a chicken or pigeon to several hundred dollars for a large elephant with a mahout. Price is mainly dependent on the size of the animal, everything else being equal. Most animals in good condition fall in the $20–$40 category.

Lineol Indians: value $40 each; fire: value $40; tents: value $150; and cowboy at the stake: value $125.

Lineol: elephant with mahout; value $250.

✦ ELASTOLIN

The O. & M. Hausser Company produced composition soldiers from 1904 to the late 1950s when, after a period of overlap when both plastic and composition figures were made, they switched completely to plastic (see page 277). Ironically they continued to call their soldiers Elastolin, the term they had invented in 1912 to signify figures made of composition. Although its molds were taken over by Preiser, plastic figures are still being issued under the name Elastolin.

After the end of World War II Hausser was forbidden to make German soldiers. They concentrated on the production of Swiss soldiers (green uniforms, black helmet), apparently as a type of generic army as these soldiers were sold worldwide. Later they made American and Austrian soldiers and finally soldiers of the German Bundeswehr. In addition to these soldiers in action, they issued parading soldiers in the uniforms of Scottish, British Line Infantry, Royal British Guards, West Point Cadets, and American Revolutionary troops. After World War II, however, the primary production was of cowboys and Indians as well as zoo and farm animals.

Identification   Most Elastolin figures are readily identified by the name Elastolin cast on the bottom of the usually oval base. Until about 1932 the main production was of 10 centimeter (4 inch) and 6 centimeter (2⅜ inch) figures. Although marching figures are common, a bewildering variety of very well sculpted action figures were also made, especially of the larger size. Horses of both sizes were standing on a wooden unmarked footplate. In the early 1930s Elastolin switched through a short period of making 6.5 centimeter (2⁹⁄₁₆ inch) figures (mainly Reichwehr) to the standard and mostly collected 7 centimeter (2¾ inch) size. In the postwar period this size was maintained but most marching figures were redesigned with a longer stride and therefore more elongated base. Until the late 1930s the word Elastolin was in capital roman lettering, but after that time new molds were supplied with letters of script. Cowboys, Indians, knights, and civilians were all marked in the same fashion. Animals without a foot plate were usually unmarked, making them very difficult to distinguish, especially from Tipple Topple, which made the same animals. In the postwar period some animals were rubber stamped GERMANY on the underside.

Prices Some of the most common composition soldiers in the United States are Elastolin's 10 centimeter (4 inch) marching Americans. These are of little value, the best preserved selling for $5 to $10. Musicians and flag bearers as well as 10 cen-

Elastolin: 10-centimeter mounted New York policeman; value $150.

Some of the many different uniforms depicted by Elastolin in the 10-centimeter size; value $40 each.

Elastolin 6-centimeter, 6.5-centimeter, 7-centimeter, and 10-centimeter figures.

timeter (4 inch) figures in action are worth more but generally sell for less than $30 each even in the best of condition. Mounted figures may fetch up to $50 and horsedrawn vehicles $200 to $300. The 6 centimeter soldiers are also common. Marching figures are almost worthless and even action figures and mounted figures sell for less than $10. Horsedrawn vehicles generally sell for $75 to $100.

The 6.5 centimeter (2⁹⁄₁₆ inch) figures are also relatively cheap, usually selling for about two-thirds of the value of the 7 centimeter (2¾ inch) figures.

A sampling of prices of 7 centimeter (2¾ inch) figures in good (but not mint) condition is given below.

29/20	Hitler in brown shirt saluting	$100
30/15N	Hitler marching saluting	$150
30/3	Goebels standing saluting	$150
30/9	Hess in SS uniform saluting	$200
26/30½	Goering standing in Air Force uniform	$125

#29/20, #30/15N, #30/3, and #30/9.

#9/7, #26/30, #1/2, #14/20, #650/1, and #649.

#18/20, #25/21, and #25/406.

9/7	Goering standing in brown shirt	$100
14/20	Admiral Raeder saluting	$75
650/1	Mackensen in Hussar uniform	$100
649	Hindenburg in coat	$100
18/20	Franco saluting	$125
25/21	Mussolini walking saluting	$150
25/406	Mussolini saluting on standing horse	$300

Several personality figures were made with very accurate porcelain heads, more as souvenirs than actual toys in the late 1930s. Hitler, Goering, Franco, and Mussolini are the personalities made with these heads. They are quite rare and prices are almost twice the price of the same figure with a composition head.

0/12	Marching	$15
51	Flag bearer with flag	$75
50/12	Standing at attention	$25
60N	Standing in coat	$25
550/6	Officer standing with map	$30
552/7N	Officer standing with arm akimbo	$35
651	Officer standing with sword	$30
47/1–47/28	Marching musicians	$30 each
47/20	Fanfare trumpeter	$75
47/24	Marching with "Jingling Johnny"	$75

#0/12 and #51.

#60N, #550/6, and
#552/7N.

#47/15, #47/20,
and #47/24.

#575, #620,
and #621.

575	Throwing hand grenade	$30
620	Advancing with rifle	$25
621	Officer advancing with sword	$35
621/3	Officer running with pistol	$30
624	Lying shooting	$20
626	Kneeling shooting	$20
628	Standing shooting	$20
640	Clubbing with rifle	$30
664/9	Sitting with machine gun	$35
664/10	Lying with machine gun	$30
664/7	Lying with ammobelt	$25
664/8	Sitting with rangefinder	$30
664/48	Lying with shell and shell boxes	$45
665/1	Lying with rangefinder	$30
665/7	Standing with rangefinder	$30
401	Mounted man	$75
402	Mounted officer	$75
550/2	Lying with cigarette	$30
550/3	Lying with canister	$30

#621/3, #628, and #640.

#664/9, #664/8, and #665/7.

#401 and #402.

#550/20, #550/25,
#550/26, and
#550/28.

#652/6, #656/1N,
and #656/2N.

550/20	Sitting with accordion	$40
550/25	Drying himself with towel	$35
550/26	Washing hair	$35
550/28	Shaving	$35
652/4	Lying wounded	$25
652/6	Falling shot	$35
656/1N	Nurse kneeling	$25
656/2N	Nurse standing	$25

#656/11,
#656/22, and
#657/10.

#659/13 and
#659/15.

#650/22, #664/20,
and #664/8.

656/11	Stretcher bearers with stretcher	$75
656/22	Medic walking	$35
657/10	Doctor	$30
659/13	Sitting with telephone	$35
659/15	Double figure sitting with radio	$100
650/22	Standing with searchlight	$35
664/17	Kneeling with shell case	$30
664/23	Standing with shell	$35
664/20	Standing with periscope	$30

#586.

#14/12.

664/8	Large explosion	$40
662/32	Rubber boat with two rowing	$150
586	Flamethrower	$50
14/12	Navy marching	$30

Postwar Swiss, American, and Bundeswehr figures are generally cheaper, the price being about two-thirds of the price of the equivalent prewar figure of a German soldier.

Elastolin cowboys and Indians are very colorful, animated figures. A much greater variety was made compared to Lineol. Foot figures are generally priced from $15 to $25 and mounted $35 to $40. The complete prairie wagon (two horses, rider, kneeling shooting man) in good condition costs about $200.

Elastolin knights [which are 6.5 centimeters (2⁹⁄₁₆ inches)] cost $20 to $30 for foot figures, $30 to $40 for mounted. American revolutionary figures cost $25 to $35. Scottish marching men cost about $40 to $50.

Elastolin animals are more common than Lineol animals. The price therefore tends to be lower, varying from $5 to $20 for small- to medium-size animals and $25 to $75 for larger animals.

Postwar Germans (Bundeswehr): #047/12, #051, #664/17, and #626. Value of flag: $45; other $15.

Elastolin: Indians; value $25 each.

Elastolin stage coach: value $650; and mounted Indians: value $40.

Elastolin prairie wagon; value $200.

Elastolin mounted knight; value $40.

✤ PFEIFER/TIPPLE-TOPPLE

The early Pfeifer figures are among the most attractive of all composition figures. Although generally not in very exciting positions, the painting is exquisite. They are readily recognized by their lack of foot plate and rather large size (11.5 centimeters—4¼ inches). Except for the cowboys and Indians these figures predate the Great War. The value for good foot figures is about $40. Mounted figures sell for about $75 to $100.

Tipple-Topple is the name used for the later figures produced in the 7 centimeter (2¾ inch) size. After the takeover of the company by Hausser in 1925 the production was closely coordinated with Hausser's Elastolin figures and a certain exchange of molds appears to have taken place. Many figures are identical to Elastolin figures except that the foot plate is blank or, in the case of postwar animals, stamped with the word AUSTRIA. Some figures have the word TIPPLE-TOPPLE molded on the underside of the base. However, both Tipple-Topple and Elastolin continued independent production of certain figures.

*Pfeifer Royal Danish
Guard. Value of flag
bearer: $85; others
$50 each.*

*Pfeifer U.S. soldiers;
value $50 each.*

Generally Tipple-Topple's painting is of slightly lower quality than the equivalent Elastolin figure. The value is, for soldiers, cowboys, and Indians, usually half to two-thirds of the equivalent Elastolin figure, whereas animals tend to be the same as they can be extremely difficult if not impossible to tell apart.

Tipple-Topple ceased producing composition figures in the 1950s.

✤ KRESGE

The so-called Kresge soldiers are 7 centimeter (2¾ inch) composition figures of English redcoats in uniforms from the early nineteenth century. These figures were made in Germany before the second World War. The figures are unmarked and the maker is unknown. A round paper label still remaining under some of

*Kresge figures.
Value of flag:
$30; others $10.*

Kresge figures: value $10 each.

*Kresge: Wellington;
value $40.*

them has the name Kresge on it, referring to the chain of stores in which they were sold.

Kresge soldiers in good condition sell from $5 to $15.

✦ LEYLA

One of the more successful companies which competed with Elastolin and Lineol was the German company Leyla. In the pre-war period it made a selection of German soldiers marching and in action. After the war it made U.S. soldiers but concentrated most of its production on cowboys and Indians.

Leyla: two prewar and one postwar figure; value $8 each.

Leyla: prewar soldiers; value $8 each.

Leyla figures usually have square bases with the name embossed underneath. They are considerably more crude than those made by Elastolin and Lineol and sell for $2 to $8.

✤ Durso

The Belgian company Durso made a large variety of 7 centimeter (2¾ inch) composition soldiers both before and after World War II. Several of them are excellent, both in sculpture and painting. Best known are their portrait figures of the victors of World War II, Montgomery, Eisenhower, and Stalin being the most common. These portrait figures sell for $10 to $30, other figures for $5 to $15.

Durso personality figures of Stalin and Zhukov (sometimes misidentified as Tito); value $35 each.

Various Durso soldiers; value $15 each.

Durso World War II Belgian soldiers; value $15 each.

Durso Arab on camel; value $40.

✤ ITALIAN COMPOSITION FIGURES

Before, and especially after World War II, several Italian companies made composition figures. Best known are Brevetta and Chialu. The latter company made a number of American Civil War figures, ignored by other makers, as well as cowboys and Indians. Their figures are sometimes in very imaginative positions and the colors tend to be bright if not outright garish. Most figures are marked either on the bottom or on the side of the base. Chialu figures sell for $5 to $15, mounted figures $15 to $25.

Figures by Brevetta; value $15 each.

Figures by Chialu. Value of paratrooper $30; other $15 each.

✤ TRICO

Composition figures, usually copies of Elastolin or Lineol figures, were also made in prewar Japan. Best known are the figures made by the company Trico. These are rather clumsy, poorly painted figures in various sizes. TRICO usually is embossed on the underside of the base. The price is low, $2–$6 for these quite unattractive soldiers.

Trico soldiers and railroad figures; value less than $10 each.

Trico 11-centimeter
figures; value $5 each.

Trico 6-centimeter
soldiers; value $5
each.

♣ OTHER COMPOSITION MAKERS

A large number of companies produced composition figures
before (and some after) World War II, especially in Germany.
Generally these figures are of poor quality, regarding both the
sculpture and painting. A few like Kienel of Germany attained a
fairly high quality. Kienel figures are unmarked but can be recognized by the healthy farm boy appearance of their soldiers.

The price for unmarked figures tends to be low, $2 to
$15, unless of unusual quality or subject.

Kienel German soldiers;
value $10 each.

Kienel German
soldiers; value
$10 to $15 each.

Napoleonic soldiers by an
unknown French manufacturer;
value $20 each.

10-centimeter composition soldiers
by the French dollmaker S.F.B.J.;
value $40 each.

Danolin figures
made in Denmark
immediately after
World War II;
value $10 each.

Composition soldiers made in Mexico; value $5 each.

✤ AMERICAN COMPOSITION

Although the process for making highly detailed and durable figures of composition was perfected in continental Europe by the turn of the century, it was never mastered in the United States. The one exception was a company named Schoenhut, which produced a 4½ inch (113 millimeter) standing American soldier in 1903. Schoenhut is most famous for its Humpty Dumpty circus.

The outbreak of World War II and the subsequent restriction on the use of metals led to a belated effort to produce composition soldiers in the United States. Existing dime-store companies such as Auburn (see page 310), Manoil (see page 182) and Historical Miniatures (see page 77) experimented with the material. Auburn never produced composition soldiers commercially. Manoil made a few readily biodegradable figures for less than a year. Historical Miniatures was able to produce a few halfway decent GIs and personality figures in very small numbers.

Two companies, Molded Products and Playwood Plastics, however, specialized in composition figures sculpted in the dime-store tradition and, in spite of the crudity of their products, met with commercial success. They were—besides Beton's plastic figures (see page 279)—the only game in town. At the end of the war they promptly went under!

MOLDED PRODUCTS   This company was the result of foresighted calculation by its founders, Leslie S. Steinam and his son. Predicting metal shortages in November 1941, they incorporat-

Molded Products: soldiers of dime-store size; value $10 each.

ed the new company to produce composition dime-store figures. As metal soldiers disappeared, grossly inferior molded products took their place on the shelves of the five- and ten-cent stores only to disappear as fast as they had popped up when metal again became available. However, the crude soldiers had served their purpose as common fodder for eager child warriors.

Identification   Molded Products soldiers (and animals) are easily recognized by the round hole in the bottom (they were placed on a nail in a wooden board when dried). They are all of the 3¼ inch (84 millimeter) size with either World War I or World War II style helmets.

Prices   Unattractive almost in the extreme, Molded Products figures are collectibles, although they command a fairly low price. The prices below refer to good, but not mint, figures.

Soldier with open parachute	$10
Pilot	$10
Soldier in gasmask	$10
Lying with machine gun	$10
Sitting with anti-aircraft gun	$10
Marching	$5
Marching with flag	$10
Marching sailor	$5

Marching marine $10
Mounted soldier $20
Cowboy $5
Indian $5
Horse $2
Cow $2
Pig $2

PLAYWOOD PLASTICS   Prior to World War II Gold Metal, a subsidiary of Transogram, was making trenches, bunkers, and the like out of wood and papier-mâché for use with dime-store figures as well as buildings for nativity scenes. Playwood Plastics, another subsidiary of Transogram, started production of composition dime-store figures in 1942 or 1943. As its only competition were the equally poorly executed Molded Products figures and the somewhat smaller Beton plastic figures (see page 186). It was successful until the end of the war when soldier production ceased.

Identification   Playwood Plastics are readily recognized by the trademark, a capital P in a triangle, marked on all its figures. The numbers of the figures (all in the 400s) can also be found on the figures although they are often difficult to read because of the poor material used. The figures are all in the classical 3¼ inch (84 millimeter) dime-store scale and have World War II-style pot helmets.

Playwood Plastics: soldiers in dime-store size; value $10 each.

Prices Remarkably, these crude figures, like their Molded Products brethren, are collected and saleable (see list below). Golden Metal-Transogram bunkers and trenches sell for $30 to $40 each. The prices below refer to figures in good, but not mint, condition.

401	Marching GI	$5
402	Flag bearer	$10
403	Kneeling machine gunner	$10
404	Stretcher bearer	$10
405	Anti-aircraft gun with operator	$15
406	Soldier in gasmask	$10
408	Lying machine gunner	$10
409	GI advancing with tommy gun	$10
410	Cannon with gunner	$10
411	GI in forage cap with slung rifle	$10
412	Paratrooper	$10
413	Seated machine gunner	$10
414	Kneeling firing	$10
415	Advancing with rifle	$30

HISTORICAL MINIATURES True to the high quality of its lead soldiers (see page 77), Historical Miniatures made the best composition figures produced in the United States. They are still vastly inferior to their European counterparts but at least are not outright ugly.

Identification Historical Miniatures did not mark their figures, making identification problematic. However, they made only a few poses of GIs (see below) and a number of personalities. Both sculpting and painting are greatly superior to the other U.S. composition figures. The GIs are in greenish uniforms with World War II-style pot helmets.

Prices Quite rare and of relatively good quality, Historical Miniatures command fairly high prices. The prices listed below refer to good, but not mint, figures.

Historical Miniatures: GIs attacking; value $25 each.

Historical Miniatures: GIs lying firing; value $25 each.

Churchill	$40
MacArthur	$40
Roosevelt	$40
Montgomery	$40
GI attacking with rifle	$25
GI attacking with flag	$50
Officer attacking with pistol	$25
GI kneeling firing	$25
GI lying firing	$25

IX

Ceramic

❧ Ceramic soldiers are highly impractical as toys because of their fragility. Indeed only the Japanese seem to have made them in any number. These Japanese ceramic soldiers seem to have enjoyed a certain degree of success in the United States as they show up quite frequently. The manufacturers are not known as the figures are only marked JAPAN or MADE IN JAPAN. Most are 2 to 3 inches (50 to 76 millimeters) in height and very clumsy looking. They sell for $10 to $25 each, depending on condition and the subject depicted.

Ceramics were also used by the Japanese to make small buildings and I have seen small ceramic vehicle recognition models made in Germany during World War II.

Only a few collectors are attracted by soldiers and/or equipment made in this material, which sell for a few dollars each.

Japanese-made ceramic soldiers; value: large $20; small $10 each.

Japanese-made ceramic soldiers; value $10 each.

X.

Celluloid

✤ This material, forerunner of plastic, found many uses before the advent of modern plastics. Celluloid was invented in the 1870s. Hollow celluloid toys were produced in several countries including the United States but became especially popular in Japan. Soldiers as well as cowboys and Indians and animals were produced in large numbers in Japan in the 1930s and immediately after World War II. The soldiers are usually between 2 inches (50 millimeters) and 3 inches (75 millimeters) large. They are very clumsy looking, highly flammable, and quite brittle, and find little appeal among today's toy soldier collectors. No price level has been established.

XI

Plastic

✣ As a material for toy soldiers plastic is close to ideal: It is cheap, light, nontoxic, and is able to be cast with the greatest detail. Its drawbacks include some difficulty in getting paint to adhere properly and a tendency for thin parts (rifles, swords) to break or warp. Nevertheless, plastic has almost completely taken over from such time-honored materials as lead and composition. Plastic was invented in the 1930s but was slow to catch on. Some companies, like Hausser in Germany, experimented with the material in the late 1930s. Jones in the United States turned it down because of the expense of the machinery. Hausser actually produced some small plastic figures during the war and plastic was used by "Wiking" to produce identification models of airplanes in Germany. Major production, however, started in the

United States in 1944 when Beton introduced its soldiers to compete with composition dime-stores figures, and, in England, in 1946 when Malleable Moldings introduced painted plastic figures of the 54 millimeter (2⅛ inch) size. In Denmark, Reisler started making plastic figures in 1949 to take up the void created by the loss of Heyde. Whereas Malleable Molding had but a short life, its competitor in England, Herald, did well and was soon taken over by Britains, which seemed to have seen the handwriting on the wall. In the 1950s many companies such as Britains, Timpo, Cherilea and Crescent in England, and Hausser in Germany switched to making their soldiers in plastic. Most companies that did not switch simply did not survive. Soon after the war Starlux of France started producing plastic soldiers of high quality. This production continues to this day. In many other countries small companies produced plastic soldiers of varying but usually quite poor quality.

At the present time early plastic soldiers, which are generally of very high quality compared to most of today's output, are becoming collectibles as the generation that played with them reaches the age for nostalgia.

Hausser (Elastolin) armored halftruck made of plastic during World War II as a winter help premium; value $30.

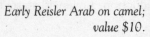

Early Reisler Arab on camel; value $10.

Reisler: Royal Danish Guard in guardhouse, a favorite Copenhagen souvenir since about 1950; value $10.

Reisler: early Danish soldiers; value $5 each.

✤ BETON

Bergen Toy & Novelty Co. (Beton) started its career in 1935 in New Jersey. Initially it made lead soldiers, but as early as 1938 it started issuing plastic soldiers, being the first company to do so. Its plastic soldiers issued during the war made their way into millions of American homes through eager wargaming boys. Together with Molded Products and Playwood Plastics, Beton (a contraction of its full name) completely dominated the dime-store soldier market in the war years.

Beton's figures were smaller than the standard dime-store figures [2¾ inches (70 millimeters) rather than 3¼ inches (52 millimeters)] making them less compatible with already existing childhood armies than were the competition's clumsy composition warriors.

Production continued after the war but competition, first from renewed production of lead dime-store figures and later from much cheaper Japanese plastic soldiers, forced the company out of business around 1958.

Beton: unpainted soldiers; value $4 each.

Beton: minimally painted soldiers; value $5 each.

Identification Beton figures are all marked under the base, BETON, MADE IN U.S.A. The soldiers came both with face and hands painted and unpainted.

Prices The price for Beton figures is quite low. Foot figures cost $3 to $5 each, mounted figures $5 to $10. Boxed sets are worth about 25 percent more than the sum of the individual pieces.

✦ MARX PLAY SETS

Starting in 1921 the Marx company made a large variety of toys, becoming best known for its tin wind-ups. By the 1950s it had become the world's biggest toy manufacturer but after a couple of changes of ownership in the 1970s, the company finally went bankrupt in 1980. An important aspect of Marx's production after World War II was its playsets, a mixture of tin plate or plastic buildings and plastic figures. As most of the molds still exist,

Marx: large-size soldiers; value $10 each.

Marx: large-size Indians; value $10 each.

a number of figures are now being reissued. Marx's plastic figures vary in size from 6 inch (152 millimeter) soldiers in action to 2⅛ inch (54 millimeter) figures. Most are 2⅜ inches (60 millimeters) in size. Most are unpainted, but some, such as the soldiers of the "Warriors of the World" set, are painted.

Identification   Many figures are marked LOUMAR.

Prices   Because of the reissue from the original molds of many Marx figures, it is difficult to price individual figures, but generally they, with few exceptions, cost only a few dollars.

The following list gives the price of some complete boxed sets (not necessarily mint, but with all parts extant). Several variations of most sets exist.

Medieval Castle	$300
Warriors of the World	$250
Ancient Chinese Warriors (individually boxed)	$35

Alamo	$350
Robin Hood Castle	$300
Battle of the Blue and Grey (Civil War)	$550
U.S. Army Training center	$200
Battleground (World War II)	$200
Beachhead landing (World War II)	$150
Daniel Boone Frontier	$500
Fort Apache	$150
Lone Ranger Ranch	$250
Roy Rogers Ranch	$175
Modern Farm	$150
Noah's Ark	$200

Marx: *Warriors of the World boxed set; value $75.*

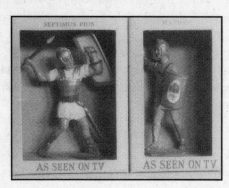

Marx: *Warriors of the World; value $10 each.*

Marx: "Babes in Toyland" boxed set; value $100.

Marx: "Babes in Toyland"; value $12 each.

Marx: General Pickett; value $10.

✤ PECO

Among the multitude of relatively small and/or short-lived U.S. plastic soldier companies, PECO (Product Engineering Company) stands out for the quality and imagination of its products. PECO of Portland, Oregon, produced soldiers only from 1952 to 1956. PECO's soldiers, cowboys, Indians, and pirates were all of 3½ inch (88 millimeter) size. The sculpting was excellent but what really set them apart were the separately cast pieces of equipment and head gear. Smaller and cheaper figures incorporating some of the features soon forced closure of PECO's soldier production.

Identification PECO was the only company to produce plastic soldiers with detachable parts in the 3½ inch (88 millimeter) size.

Prices PECO figures rank among the more expensive plastic soldiers. With all equipment soldiers cost $10 to $15, and pirates $15 to $20. A figure of Davey Crockett equipped with coon cap, powder horn, musket, pistol, and bread pouch is the most sought after, costing about $30.

♣ OTHER AMERICAN MAKERS OF PLASTIC SOLDIERS

A large number of short-lived and/or small U. S. companies have made various plastic soldiers and figures. Makers such as Plastic Toys Inc., Ausley, Ajax, Lido, and Thomas Toys were among the earliest producers of plastic soldiers, but all fell victim to competition from products of larger U. S. companies (e.g., Marx) or foreign companies which could make cheaper items. The soldiers they produced are usually marked with the company name and most sell for a few dollars each at most.

♣ HASBRO'S GI JOE

The first successful action figure with movable limbs was Hasbro's GI Joe, which hit the toy stores in 1964 and was an immediate success. Different types in various World War II outfits were made, both of U.S. Army and foreign armies. In 1967 Vietnam War outfits were added but the anti-war movement of the time led Hasbro to change the title and uniforms to Adventurers. These found little favor and production stopped by 1975.

GI Joes are the same height as the earlier and much more successful Barbie doll, 11½ inches (295 millimeters).

Identification It is not easy to misidentify this large action figure. Until 1970 the figures had hair painted on. After 1970 flock hair was used (and later moustaches and beards). Early U.S. Army figures have a facial scar.

Hasbro's GI Joe;
value $100.

Prices   The list below refers to early boxed, complete sets. GI Joe Adventurers sell for about $30 each.

7500	GI Joe action soldier	$100
7501	Combat set A	$40
7502	Combat set B	$40
7512	Bivouac set A	$10
7517	Command post set	$10
7521	Military police set	$50
7531	Ski patrol set	$50
7537	West Point Cadet set	$125
7600	GI Joe action sailor	$200
7602	Frogman set	$125
7612	Shore patrol set	$50
7624	Annapolis Cadet set	$200
7700	GI Joe action marine	$100
7710	Dress parade set	$35
7719	Medic set	$50
7800	GI Joe action pilot	$200
7801	Survival set	$60
7803	Dress uniform set	$35
8100	German Storm Trooper	$200
8104	British commando	$200

| 8300 | Equipment for German storm trooper | $75 |
| 8304 | Eqipment for British commando | $75 |

For greater detail see Killian & Griffin, *G.I. Joe Collectibles*.

♣ BRITAINS

HERALD   The English company Zang was among the first to produce plastic soldiers. These were made of hard and rather brittle plastic, which was then painted. The company became part of Herald in about 1954 and expansion of the production continued. In 1959 Herald was bought by Britains (if you can't beat them, buy them!) and the series was continued until 1975. Until 1966 Herald figures were made in England but from then onward in Hong Kong. The Hong Kong figures were made in the same molds but slightly softer and much less brittle plastic was used. Unfortunately, the paint work was also less subtle than the earlier ones. They were sold both in boxes and in bulk.

Identification   All figures are marked either with the Zang symbol, a stylized herald in a circle, HERALD, and/or BRITAINS, as well as MADE IN ENGLAND or MADE IN HONG KONG. The foot figures are of the standard 54 millimeter (2⅛ inch) size.

Britains Herald: Greek warriors; value $10 each.

Boxed set of
Britains Herald
Scottish troops;
value $125.

Britains Herald: mounted
Greek warrior; value $20.

Britains Herald: Life Guard;
value $15.

Britains Herald: Life
Guards; value $6 each.

Britains Herald:
Life Guards;
value $15 each.

Britains Herald: cow-
boys; value $8 each.

Most are on a rounded plastic base. Mounted figures are baseless. Horses are marked on the belly. Many feel that these early plastic figures are among the most beautifully crafted toy soldiers.

Prices Britains' Heralds came in boxed sets of different numbers of figures. These vary in price from $50 to $200, depending on subject and size. Single figures sell for about $10 for foot figures and $20 for mounted figures.

SWOPPET FIGURES Herald introduced these figures with interchangeable parts in 1958 and Britains took over this series the following year. They remained in production until the late 1970s.

Identification Other companies, notably Timpo and Hausser, produced soldiers with interchangeable parts. Britains are however generally better proportioned and in more natural

Box of Swoppets; value $100.

Swoppet: knight;
value $50.

Swoppet: knight;
value $50.

Swoppet: knight;
value $50.

Swoppet: knights; value $20 each.

Swoppet: archers; value $15 each.

Swoppet: Union cavalrist with banner; value $40.

positions, especially when compared with Timpo's. They are clearly marked on the base.

Prices   Swoppet figures are rarely found in boxed sets, the prices of which range from $100 to $200. Foot figures, if complete, cost about $15, mounted figures $40 to $60. It is important to assure that all pieces are present as each figure or the absence of even small parts reduces the price drastically.

EYES RIGHT This series of excellent plastic soldiers was brought out by Britains in the 1960s to replace the best metal figures. Only marching and band figures (mounted and foot) were produced. The series was rather short lived but its figures are among the most sought-after plastic soldiers.

Identification Eyes Right figures have movable heads and arms. They are clearly marked with BRITAINS LTD. and MADE IN ENGLAND. They are all of the standard 54 millimeter (2⅛ inch) size.

Prices Eyes Right sets range in size from the single mounted figure of the Royal Canadian Mounted Policeman to bands of twenty figures. The price for boxed sets varies from $30 to $500. Individual figures vary in price from $10 (foot) to $30 (mounted).

Eyes Right: guard band in original box; value $40.

Eyes Right: Life Guard band figures; value $20 each.

DEETAIL This series was introduced in 1971 and has been very successful, both when sold in boxed sets and in bulk. Vehicles and groups with mortars and machine guns have also been produced.

  Identification Plastic soldiers on metal bases, these figures are instantly recognized and the boxes are clearly marked.

Deetail: Japanese and English soldiers; value $5 each.

Deetail: English and German soldiers from the North African campaign; value $5 each.

Deetail: German mortar team; value $30.

Deetail: German halftrack motorcycle; value $50.

Deetail: U.S. jeep; value $35.

Deetail: British scout car; value $45.

Prices   Boxed sets of Deetail figures range from $25 to $150 depending on size and subject. The most expensive sets are of the Foreign Legion and Arabs as well as the Waterloo series. Individual foot figures range from $5 to $20, composite figures of, for instance, machine guns, from $30 to $50, and mounted figures from $10 to $30.

✤ ELASTOLIN

Hausser, the parent company of Elastolin, first experimented with plastic in the late 1930s. No production of soldiers got under way at that time but during the war the company produced a series of small tokens in plastic which were sold for the benefit of the WHW (the Winter Relief Charity). It was not until the mid-1950s that "real" Elastolin plastic figures were produced. The first was an ambitious project of a stagecoach drawn by four horses. Soon thereafter followed cowboys and Indians, Vikings, Romans, Mongols, Turks, Normans, Landsknechts, Prussian eighteenth century infantry, American Revolutionary Colonials and their British Redcoat opponents, all in the 70 millimeter (2¾ inch) scale. A series of magnificent medieval siege equipment pieces were produced with these figures. Figures of 40 millimeter (1⅝ inch) scale were also produced in great quantities and with the same subjects as the 70 millimeter (2¾ inch) figures.

All Elastolin's figures were hand painted, but in the late 1970s the quality of painting deteriorated and, by the time Hausser closed down in 1986, was quite poor. The molds were taken over by the Preiser Company and production, still using the name Elastolin, resumed but the painting was now of considerably higher quality. This production is continuing and new items are still being added to the line by Preiser.

Identification   All figures are marked ELASTOLIN on the underside of the base.

In the early figures the skin color was painted (the plastic used was white), whereas later figures were made in a rather ugly pinkish skin color so bare parts on the figures were left unpainted.

Prices   The earliest Hausser Elastolin figures command the highest prices.

Elastolin: 40-millimeter Roman chariot; value $75.

The list below gives a sampling of prices for some of the multitude of figures made in plastic by Elastolin. Note that many of these are still being made by Preiser.

Cowboys

Tied to tree	$20
Clubbing	$12
Kneeling firing rifle	$12
Kneeling firing pistol	$12
Standing firing rifle	$12
Sheriff standing with revolver	$10
Standing with lasso	$12
Bandit running with revolver and money bag	$15
Bandit running with two revolvers	$15
Bandit surrendering	$15
On galloping horse firing rifle	$20
On galloping horse holding rifle	$20
On galloping horse firing rifle	$20
On galloping horse throwing lasso	$20
On standing horse scouting	$20
On standing horse waving hat	$25
On standing horse RCMP	$35
Stagecoach with four horses	$350
Prairie Wagon with two horses	$150

Elastolin cowboys.

Elastolin mounted cowboy.

Indians

Totem Pole	$10
Standing with spear	$12
Medicine man dancing	$15
Sitting with spear	$12
Sitting with pipe	$12
Sitting with bow	$12
Squaw sitting with papoose	$10
Squaw sitting with bowl	$10
Girl with jar	$10
Standing with bow	$12
Standing with rifle	$12
Kneeling with bow	$12
Lying behind rock with rifle	$12
Lying with rifle	$12
Crawling	$12
Throwing spear	$12
Running with spear	$15
On galloping horse firing rifle	$25
On galloping horse with bow	$25
On galloping horse with spear and shield	$25
On galloping horse with tomahawk and shield	$25

*Elastolin
Indians.*

*Elastolin mounted
Indian.*

*Elastolin mounted
Indian.*

7th Cavalry

Kneeling firing	$12
Running with sword and revolver	$12
Standing with revolver and pennant	$15
On galloping horse with revolver	$25
On standing horse with pennant	$30

Elastolin 7th Cavalry.

Romans

Marching with sword	$12
Marching with trumpet	$15
Marching with standard	$18
Running with sword	$12
Throwing spear	$12
Kneeling throwing spear	$12
Fighting with sword	$12
Archer running with bow	$12
Archer firing bow	$12
On galloping horse with sword	$25
Chariot with four horses	$200

Elastolin Roman Legionaries.

Elastolin Roman Legionaries.

Vikings

Throwing spear	$8
Drawing sword	$8
Fighting with sword	$8
Fighting with axe	$8
Holding spear	$8

Mongols (all mounted on galloping horses)

With war drums	$35
With bow	$35
With spear	$35
With bugle and sword	$35
With shield and sword	$35

Turkish Janissaries

With sword	$20
With halberd	$20
With bow	$20
With sword	$20
With blunderbuss	$20

Normans

With sword	$10
With axe	$10
With spear	$10

Elastolin medieval warrior.

Mounted with lance	$25
Mounted with sword	$25
Mounted with axe	$25
Mounted with spear	$25

Landsknechts

Marching with flag	$15
Marching with halberd	$12
Marching with pipe	$12
Marching with drum	$12
Firing blunderbuss	$15
Running with flag	$15
Running with halberd	$12
On standing horse with trumpet	$30
On standing horse with standard	$30
On standing horse with spear	$25

Prussian 18th Century Infantry

Frederick the Great with two hunting dogs	$15
Frederick the Great mounted	$25
Marching with musket	$10
Marching with drum	$12
Marching with pipe	$12
Marching with flag	$15
Loading musket	$12

Kneeling firing	$10
Standing firing	$10

American Revolutionary Colonials and Redcoats

George Washington standing	$10
George Washington mounted	$20
Marching with musket	$10
Marching with drum	$12
Marching with pipe	$12
Marching with flag	$15
"Spirit of '76" (3 pieces)	$40
Officer with sword	$12
Loading musket	$12
Kneeling firing	$12
Standing firing	$12

Siege Equipment

Siege tower	$135
Arrow catapult	$35

Elastolin U.S. Revolutionary soldiers.

Elastolin "Spirit of '76."

Elastolin Germans in action.

Elastolin Germans in action.

Large catapult	$75
Battering ram	$100
Howitzer with brass barrel	$50
Field cannon with brass barrel	$50
Mortar	$35

Elastolin 40 millimeter (1⅝ inch) figures cost about one half of the price for the equivalent 70 millimeter (2¾ inch) figures.

✤ STARLUX

This French company, which had produced composition soldiers before World War II, started producing painted plastic soldiers shortly after the end of the war. The quality of these 60 millimeter (2⅜ inch) figures was, and still is, very high. Although distribution in the United States has been quite poor, there has been an increasing recognition of the quality of these figures. Many are marketed in individual boxes with a cellophane front.

Identification Most Starlux figures are marked with the name on the base.

Starlux: soldiers in action; value $2 each.

Starlux: soldiers in action; value $2 each.

Starlux: French Foreign Legion; value $2 each.

Starlux: French Foreign Legion; value $2 each.

Prices Starlux has an ongoing production of a very large variety of figures. The prices below are therefore for first grade, mint, and boxed figures. Lesser grades of painted figures are also made.

Gauls and Romans foot figures	$6
Gauls and Romans mounted figures	$12
Knights on foot	$5
Knights mounted	$10
French Revolution, personality figures on wooden box	$18
French Empire foot figures	$12
French Empire mounted figures	$18
World War I figures (French and German)	$6
World War II figures (French and German)	$6
U.S. Army in action (non-boxed)	$2
French Foreign Legion in action (non-boxed)	$2

Starlux: World War II Russians; value $6 each.

Starlux: World War II Russians; value $6 each.

French Parachute Regiment in action (non-boxed)	$2
German Army in action (non-boxed)	$2
Italian Bersaglieri in action (non-boxed)	$2
Cowboys on foot (non-boxed)	$2
Cowboys mounted (non-boxed)	$3
Indians on foot	$2
Indians mounted	$3
Union infantry	$2
Union cavalry	$3
Confederate infantry	$2
Confederate cavalry	$3
Stagecoach	$25

Starlux: Arabs; value $3 each.

Starlux: Napoleonic figures; value $12 each.

Starlux: Napoleon; value $18.

Starlux: Napoleonic officer; value $18.

✤ TIMPO

Of the many makers of lead soldiers that switched to plastic Timpo was among the more successful. At the change over from hollow cast to plastic in 1956, Timpo used its earlier figures as models. Later they added Romans, Robin Hood figures, Arabs, French Foreign Legion, and still others. These painted figures are by no means as attractive as Timpo's hollow casts but are still of high quality. Swoppet-like soldiers were also made by Timpo in the late 1960s.

Identification   Most of Timpo's 54 millimeter (2⅛ inch) figures are marked with the name and/or MADE IN ENGLAND. The sturdy appearance is characteristic.

Prices   Early painted Timpo plastic figures are quite rare to find in good condition because of the increasing brittleness of the plastic as it ages. They sell from $5 to $20 each. Later figures sell for $5 to $10 each. Boxed sets cost about 25 percent more than the sum of the individual pieces.

Timpo Swoppet box.

Timpo Swoppet soldier value $8 each.

Timpo: mounted Arab; value $20.

♣ OTHER EUROPEAN PLASTIC SOLDIER MAKERS

Several of the lead soldier manufacturers tried to convert to plastic in the 1950s. Other companies were started in the 1950s and 1960s. They met with varying success, most closing down after only a few years.

Charbens (England) closed down in 1968 after having produced some quite acceptable sets, of pirates for instance, and a safari set.

Cherilea (England) made a number of successful plastic sets such as Cleopatra and her court and United Nations soldiers in action. The company still exists but produces very little and that only sporadically.

Crescent (England) also made plastic figures of some note, especially knights, both swoppet-like and regular. After a peak in the 1960s, production has gradually petered out.

Atlantic (Italy) started in 1972 lasting until the late 1980s. During its relatively short life, the company produced a vast number of unpainted but very well molded figures sold as parts of dioramalike scenes.

Reamsa (Spain) made 60-millimeter (2⅜-inch) plastic figures, both painted and unpainted. Notable are sets of famous bullfighters and cowboys and Indians. Started in 1955, the company closed in the 1980s.

Several companies in Russia and Eastern Europe, Poland in particular, also made plastic soldiers, often of surprisingly high quality. These companies are barely known in the West and have attracted few, if any, collectors.

Crescent box.

Crescent: cowboys; value $5 each.

Polish mounted officer by an unknown Polish manufacturer; value $8.

XII

Rubber

✤ This material has not been used much for producing toy soldiers—and for several good reasons. Rubber tends to age poorly as it warps and cracks. Because of its flexibility it holds the paint poorly and it is virtually impossible to show details in the casting. Nevertheless the material was used successfully by Auburn from the mid-1930s to the 1950s, and after the war by several East German companies. Like most other materials rubber was replaced by plastic in the 1950s.

✤ AUBURN

This company, which took its name from the town in Illinois where it was located, started in 1910 making automobile tires and tubes. In the 1920s it expanded into other rubber products

and in 1935 threw itself into the booming field of dime-store figures. In spite of the quite clumsy appearance of its figures the
company was very successful, not least because of its advanced
marketing techniques. In its advertising Auburn emphasized the
safety of its products. They contained no toxic dyes (before
painting the surface was treated with a vegetable-based varnish)
or other material (concern over the toxicity of lead figures started in the mid-1930s). They didn't scratch floors or break teeth
and were virtually indestructible. Packaging was also imaginative; for instance, its prize-winning set of eighteen baseball players released in 1940. During the war production of toys stopped
but was resumed again afterward. In the early 1950s Auburn con-
verted to vinyl plastic. When the company was sold and moved
to New Mexico in 1959 the production of plastic figures continued, but the company was supposedly fleeced by labor racketeers
and closed down in 1969.

Identification   Auburn is the only United States company that made rubber figures. Old rubber hardens but is still recognized as such without difficulties.

Prices   As with other toy soldiers, condition is all
important. Added to the usual paint loss is the problem that figures made of rubber seem to bend every which way and warp to
the point of toppling over.

The prices in the list that follows refer to figures in good,
but not perfect, condition.

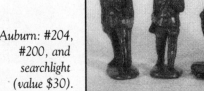

*Auburn: #204,
#200, and
searchlight
(value $30).*

200	Doughboy walking with rifle	$10
202	Doughboy with bugle	$15
204	Officer	$15
206	Stretcher bearer	$25
208	Wounded	$30
216	Doughboy kneeling with binoculars	$15
222	Doughboy crawling	$60
224	Doctor	$25
226	Nurse	$25
234	Doughboy throwing hand grenade	$20
236	Doughboy with signal flags	$50
238	Doughboy charging with tommy gun	$10
240	Doughboy on motorcycle with sidecar	$45
242	Doughboy with anti-aircraft gun	$25
272	Doughboy kneeling shooting at airplane	$30

Baseball figures

250	Pitcher	$35
252	Batter	$30
254	Catcher	$25
256	Fielder	$20
258	Runner	$20

Football players

260	Lineman	$25
272	Back	$25
264	Center	$25
266	Passer	$35
268	Carrier	$30

Animals

102	Horse	$10
106	Pig	$10
108	Cow	$10
112	Hen	$5

East German rubber figures usually depict members of the East German Army. The painting is minimal and the figures clumsy. Most are about 2½ inches (65 millimeters) tall. They are not very collectible and sell for $1 to $5 each.

*East German
rubber soldier;
value $5.*

*East German
rubber soldiers;
value $5 each.*

XIII

Equipment

❧ As anybody having played with toy soldiers knows, equipment is important. Both ancient and modern armies were more than a collection of individual soldiers. An army needed to move and although feet and horses could do in a pinch, vehicles, either horsedrawn or motorized, were much to be preferred. No modern army is complete without an air force. Even on the move soldiers have to be housed, so tents are needed. They have to eat so mobile kitchens must be available. The home base had to be secure so fortifications had to be available, bunkers for modern troops, castles for men at arms of times past and forts for the western expansion of Americans. Soldiers also need weapons of larger caliber than bows and rifles, so siege equipment and cannons are absolute musts for such miniature armies as they are for real armies. Best of all, if these weapons can actually shoot, further realism is brought to the killing fields of toy soldiers.

This demand has indeed been met by toy manufacturers, partly by the toy soldier manufacturers themselves, but not infrequently by other companies generally specialized in other fields.

VEHICLES AND CANNONS

Lead soldiers, which are mainly in the 5 to 6 centimeter (2 to 2⅜ inch) scale were largely served by equipment supplied by the toy soldier manufacturers themselves. Heyde, Mignot, and Britains all made vehicles and cannons, both horsedrawn and motorized. Cannons especially were made by many other manufacturers and readily supplemented what was otherwise available.

Composition soldiers were supplied by transport and artillery by both Hausser and Lineol, but the expense of these left room for competition as did the quite poorly equipped tank corps.

Dime-store figures were supplied with vehicles and cannons that were generally unsatisfactorily small and out-of-scale, leaving room for improvement by others. On the other hand, many children did not mind mixing out-of-scale figures or equipment. A whole subdivision of the ¹⁄₄₂ scale car manufacturers consisting of military vehicles, often of extraordinary detail and realism, exists.

Tin-plate Vehicles and Cannons

Tin plate, painted or lithographed, lends itself well to the construction of medium-sized to large vehicles and cannons. Several companies made such equipment.

✤ HAUSSER (ELASTOLIN, SEE PAGE 251)

Until plastic took over in the late 1950s, Hausser made a number of excellent vehicles and cannons of an appropriate size for their composition soldiers. The larger type (10 centimeters, 4 inches) as well as the smaller (6 centimeters, 2⅜ inches) were horsedrawn only. In the early 1930s when the 7 centimeter (2¾ inch) size soldiers were introduced these were initially supplied with a number of oversized vehicles, only later scaled down to fit better. After the war some minor modifications were made and the colors were changed from the camouflage and green, light and dark brown of the early thirties and the variable shade of slate of the late thirties to a uniform dark khaki. Strangely the plastic soldiers were left without equipment and transport, except for the unsurpassed medieval siege equipment.

Identification   Early horsedrawn vehicles are unmarked and difficult to tell from Lineol's equipment. The later equipment is marked HAUSSER, either stamped in the metal or on the tires.

Prices   The most expensive Hausser vehicles are the early oversized trucks. These might cost well over $1,000 if well preserved. The cheapest are the small-scale horsedrawn vehicles fetching $100 to $150. Cannons vary in price according to size. Examples are given below.

10 centimeter (4 inch) Horsedrawn cannon	$250
10 centimeter (4 inch) Horsedrawn ambulance	$300
6 centimeter (2⅜ inch) Horsedrawn cannon	$100
6 centimeter (2⅜ inch) Horsedrawn ambulance	$125
Kübelwagen	$900
Searchlight truck	$1,200

Hausser: 10-centimeter horsedrawn cannon; value $250.

Hausser: 10-centimeter horsedrawn ambulance; value $300.

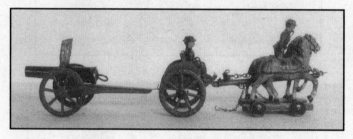

Hausser: 6-centimeter horsedrawn cannon; value $100.

Hausser: 6-centimeter horsedrawn ambulance; value $125.

Hausser: Kübelwagen with field kitchen. Value $750; field kitchen $125.

Halftrack prime mover	$3,000
Small peashooting cannon	$35
Anti-tank gun (PAK)	$75
Anti-aircraft gun (88 millimeters)	$400

Hausser: 7-centimeter horsedrawn wagon; value $250.

Hausser: Kübelwagen with PAK. Note the metal tires on car, dating vehicle to the war years; value $700; PAK $85.

Hausser: 6-wheeled truck; value $1,200.

Hausser: AA machineguns; value $600.

✤ LINEOL (SEE PAGE 242)

The vehicles and cannons made by Lineol are very similar to those by Hausser. Early motorized Lineol vehicles were as much out of scale as Hausser's, if not more so. Lineol had no postwar production of vehicles or cannons.

Identification Early horsedrawn vehicles are unmarked. Motorized vehicles and cannons either have Lineol stamped in the metal or on the tires.

Prices Lineol equipment tends to be a bit more pricy than the equivalent Hausser equipment. Examples of prices for good, but not mint, pieces are given below.

9 centimeter (3⁹⁄₁₆ inch) Horsedrawn cannon	$300
6 centimeter (2⅜ inch) Horsedrawn cannon	$125
Kübelwagen	$1,200

Lineol: 9-centimeter horsedrawn cannon; value $400.

Armored six-wheel car	$1,000
Tank	$1,000
Anti-aircraft cannon	$500
Howitzer	$600
Anti-tank gun (PAK)	$100

Lineol: Kübelwagen; value $1,200.

Lineol: 6-wheeled armored car; value $1,000.

Lineol: tank; value $1,000.

Lineol: 9-centimeter horsedrawn ambulance; value $300.

Lineol: 6-wheeled truck; value $1,200.

✣ TIPP CO.

This German tin-plate company made a number of vehicles that are perfectly compatible with both Elastolin and Lineol soldiers (7 centimeters, 2¾ inches). Although it produced civilian vehicles after the war, no military vehicles were made at that time.

Identification Tipp Co. vehicles are generally unmarked, but the wheels and tires are made of lithographed tin. Tipp Co. vehicles are smaller than the equivalent Lineol or Hausser vehicles and, as the tin plate is thinner, considerably lighter in weight.

Prices Tipp Co. vehicles are considerably lower in cost than those for the equivalent Hausser or Lineol vehicles. As opposed to both these companies, Tipp Co. made a Führerwagen, an open black Mercedes, which is very desirable. Examples of prices for good, but not mint, vehicles are given on page 322.

Führerwagen	$2,000
Kübelwagen	$500
Searchlight truck	$700
Anti-aircraft cannon on truck	$700
Halftrack prime mover	$1,000
Ambulance	$700
Anti-aircraft cannon (88 millimeters)	$200
Large Tank	$700
Armored eight-wheel car	$500

*Tipp Co.:
Kübelwagen with
PAK; value $500;
PAK $50.*

*Tipp Co.: AA
cannon on
truck; value
$700.*

*Tipp Co.: halftrack prime mover with searchlight; value $1,000;
searchlight $150.*

Tipp Co.: ambulance;
value $700.

Tipp Co.: large tank;
value $650.

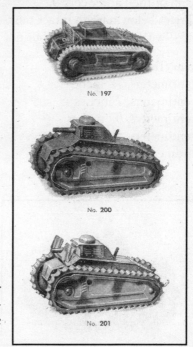

No. 197

No. 200

No. 201

Tipp Co.: smaller
tanks (from origi-
nal catalog); value
$250 to $350.

✤ GAMA

This German tin-plate manufacturer specialized in making tanks, both before and after World War II. A large variety was made; most prewar tanks have German markings while postwar tanks have American markings. Most run on rubber threads, which are often missing or broken because of age. The tin plate is lithographed.

    Identification   The word GAMA is lithographed on virtually all the tanks, usually on the front end.

    Prices   GAMA tanks are quite common and, for German lithographed toys, relatively cheap. Larger types have tools (axe and spade) attached to the fenders and sparkling machine guns under the gun turret. Tools and machine guns are often missing as is the man for the turret that is usually supplied with the larger prewar tanks (he is in black uniform with beret, the uniform of the Panzer corps).

    Prices below refer to intact tanks in good, but not mint, ,condition—with all equipment (but not the man).

Prewar Tanks

19 centimeters, 7⅜ inches	$250
18 centimeters, 7 inches	$175
16 centimeters, 6¼ inches	$125
13 centimeters, 5⅛ inches	$100
9 centimeters, 3½ inches	$50

Gama: prewar tank with man; value $300.

Gama: prewar tank; value $175.

Gama: postwar tanks; value $175, $100, $75, and $35.

Postwar Tanks

19 centimeters, 7⅜ inches	$175
18 centimeters, 7 inches	$150
16 centimeters, 6¼ inches	$100
13 centimeters, 5⅛ inches	$75
9 centimeters, 3½ inches	$35

♣ MÄRKLIN

This venerable and still existing firm which specializes in toy trains made a number of tin-plate cannons before World War II.

Identification Märklin cannons are usually marked with the firm's logo or stamped MÄRKLIN. As with their other tin-plate work, the quality is very high. Older models (pre World War I) often have brass firing mechanisms.

Prices Because of their high quality, Märklin cannons tend to be relatively expensive, especially the ones on elevated emplacements (coastal guns). Examples of prices for field can-

Märklin cannon;
value $250.

nons in good condition with intact firing mechanisms are given below.

Field Cannons

21.5 centimeters, 8½ inches	$150
16.5 centimeters, 6½ inches	$120
15 centimeters, 6 inches	$75
11.5 centimeters, 4½ inches	$50

✤ ARNOLD

This German tin-plate manufacturer is best known for its post-war Jeep with a crew of three, usually military police, made in composition. This vehicle costs about $150 with crew.

Arnold jeep;
value $150.

Arnold cannon;
value $30.

*Gescha tank;
value $200.*

*Gescha cannon;
value $60.*

✤ OTHER GERMAN COMPANIES

Companies such as Ebo, Bing, Gescha, and Hess all made cannons before World War II. These are usually marked with the company's name and/or logo. The prices are generally 25 percent less than those of the equivalent Märklin piece.

✤ MARX (SEE PAGE 280)

This large toy company made a number of military vehicles and cannons that can be used with dime-store figures in particular. These lithographed toys are usually marked with Marx's logo. Examples with prices are listed below.

Army ambulance	$200
Army staff car	$350
Sparkling, climbing, fighting tank	$300

Crescent: diecast armored car with limber and cannon; value $45.

French diecast cannon; value $20.

Astra diecast cannon (English); value $30.

Turnover tank	$125
Anti-aircraft gun	$75
Large cannon	$100

✦ OTHER U.S. COMPANIES

Other U.S. companies made tin-plate vehicles, cannons, and tanks. Structo made an 11 inch (28 centimeter) long tank on metal treads which is selling for about $300. Chein of New Jersey made lithographed 8 inch (20 centimeter) military trucks selling for about $150. Sonny made an oversized (24 inch, 60 centimeter) artillery truck which sells for about $350. Nylint made an oversized (15 inch, 38 centimeter) missile carrier (about $300). Keystone made an army truck (25 inches, 62 centimeters) which today is worth almost $1,000 and Buddy L a 27-inch (67.5 centimeter) army truck with a towed cannon worth about $200 today.

✤ JAPANESE TIN MILITARY VEHICLES

A number of Japanese battery-operated military vehicles were made in the 1950s. Most are somewhat oversized, even for dime-store figures, but might do in a pinch. Note that the marks stamped on these lithographed toys are often those of the American distributor rather than the Japanese maker. Examples of some of these vehicles and prices are given below.

Army truck with double AA cannons, 14 inches
    (35.5 centimeters), Linemar $150
AA Jeep, 10 inches (25 centimeters) "K" $125
M-4 tank, 12 inches (30.5 centimeters) Taiyo $50
M-35 tank, 8 inches (20 centimeters) HTC $75
M-81 tank, 9 inches (23 centimeters) M-T $100
M-X tank, 9 inches (23 centimeters) T-N $75

Cardboard vehicles and cannons

✤ BUILT-RITE

This manufacturer of cardboard boxes made a number of vehicles and cannons before and during World War II. These could be punched out and assembled. Sets cost from $50 to $100, depending on complexity and condition.

✤ DIE-CAST VEHICLES AND CANNONS

Makers of die-cast cars, such as Dinky, Matchbox, Solido, and Tekno have also ventured onto the military scene, but virtually all of them are greatly undersized and do not go well with toy soldiers. They form collector's items in and of themselves and are generally dealt with in association with model cars.

Britains made a number of die-cast vehicles and cannons to go with their soldiers (see page 90).

AIRPLANES

Modern armies are always supported by an Air Force and so are the miniature armies of little boys. However, even new, such airplanes tend to be expensive and as antiques they are among the most expensive accessories to armies in lead or composition.

Only a few soldier manufacturers such as Britains and Mignot made their own airplanes. Most were supplied by companies that specialized in tin-plate, cardboard, or larger die-cast toys.

Tin Plate

♣ TIPP CO.

This German company is famous for its airplanes and zeppelins. Several types of airplanes were made, both single-engined fighter planes (or dive-bombers) and multi-engined transport planes or bombers.

Identification Tipp Co. did not generally mark their toys, but some are marked TC. They are, however, quite easily recognized by the lithography, especially of the wheels and tires, which are all tin plate.

Prices Expensive when made, prices today are very high and none in even half decent shape can be had for less than $1,000 unless one is very lucky indeed.

Five-engined biplane	$2,500
One-engined biplane with machine gun	$1,200
One-engined monoplane with bombs	$1,200
One-engined monoplane with ejectable pilot	$1,500
Airship Hindenburg, 28 centimeters (11 inches)	$700
Airship Hindenburg, 43 centimeters (17 inches)	$2,000

Tipp Co.: biplane with machineguns, after original catalog; value $1,200.

Tipp Co.: monoplane with bombs, after original catalog; value $1,200.

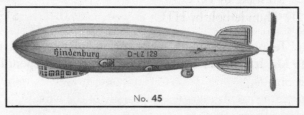

Tipp Co.: airship, length 28 centimeters, after original catalog; value $700.

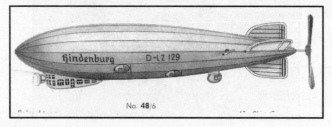

Tipp Co.: airship, length 43 centimeters, after original catalog; value $2,000.

♣ MARX (SEE PAGE 221)

This U.S. tin-plate manufacturer made a number of military air-planes in tin plate or pressed steel. These are all marked with Marx's logo.

Twin-engine U.S. Army plane	$200
Single-engine U.S. dive bomber	$200
Four-engine Flying Fortress	$250

♣ OTHER MANUFACTURERS OF TIN-PLATE AIRPLANES

Several Japanese companies made military airplanes especially in the post World War II era.

Bristol Bulldog by ScE	$150
F3F Biplane by Cragstan	$250
Spitfire by HTC	$150
P51 Mustang by HTC	$150
P-47 Thunderbolt by HTC	$150

Tekno of Denmark made an ambulance plane in the late 1930s. It sells for about $1,000 to $1,500, depending on condition.

Tekno: ambulance plane; value $1,200.

✤ AIRPLANES OF OTHER MATERIALS

Because of their size only a few full-scale airplanes to go with toy soldiers have been made in cast metal, but a few were made by Hubley (which are so marked).

Triple engine with pilot and copilot	$1,500
Spirit of St. Louis	$500

Airplanes were also made of cardboard, especially during World War II, but few of these have survived. A large number of plastic kits of airplanes to be assembled are available. These, however, form their own branch of collectibles.

CASTLES AND FORTIFICATIONS

Although a few boards hammered together make a perfectly usable castle or fortification for most armies, more elaborate structures are of course more desirable. Of the traditional lead soldier only Britains, immediately prior to World War II, made a number of now very rare buildings and fortifications (in wood and papier-maché). Composition makers produced a large number in wood, papier maché, and composition. Elastolin, especially, made beautifully detailed castles as well as more modern fortifications, trenches, and bunkers. For their cowboys they made various log cabins and palisades. The early large Elastolin castles may cost up to $800. Simple trench sections cost $25 to $50, whereas more elaborate trenches and bunkers may cost up to several hundred dollars although the price is usually lower.

After the war Elastolin made plastic castles. These vary in price from $75 to $300.

A myriad of other companies, German, English, Danish, and American among them, made castles and fortifications, the prices of which vary widely from $25 to $50 for a small fiberboard castle to $300 for Marx's castle fort of tin plate with plastic knights.

The market for castles is relatively small because of the amount of space they occupy.

Tents by various makers; value: square $20, round $10.

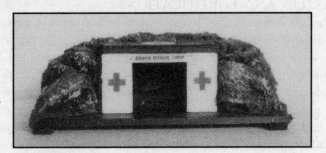

Hausser (Elastolin): advanced dressing station; value $200.

Hausser (Elastolin): trench with observation post; value $100.

Hausser (Elastolin) castle; value $500.

Danish-made castle; value $300.

Danish-made castle; value $200.

Danish-made castle;
value $100.

Danish-made castle;
value $100.

German-made castle;
value $150.

Periodicals

British Model Soldier Society
Treasurer: Ian R. Webb
35 St. John's Road
Chelmsford
Essex CM2 OTX, England
Founded 1935. Monthly
meetings in London.
Arranges shows. Issues *The Bulletin* and newsletter.

Figuren. Zeitschrift für Sammler von Aufstell-Figuren
Andreas Pietruschka
Spenerstrasse 17
1000 Berlin 1, Germany
Founded 1972. Quarterly

issues on composition and
plastic figures. In German.

Old Toy Soldier
209 North Lombard
Oak Park, IL 60302-2503
Founded 1976. Bimonthly
issues on toy soldiers.
Arranges shows.

Toy Soldier Review
Vintage Castings, Inc.
127 74th Street
North Bergen, NJ 07047
Founded 1985. Quarterly
issues on toy soldiers.
Arranges shows.

Gama: postwar tank; value $175.

Tipp Co.: small size searchlight truck; value $150.

Tipp Co.: small size ambulance; value $150.

Tipp Co.: small size AA truck; value $150.

Auction Houses

♣ Some auction houses have special toy soldier auctions. These are listed below. Many others have toy auctions at which toy soldiers may be for sale, as is the case with many estate auctions.

Christies
85 Old Brompton Road
London SW 7 3LD, England
Tel: 011,44-71-581-7611
Fax: 011,44-71-321-3321

Henry Kurtz Ltd.
163 Amsterdam Avenue,
Suite 136
New York, NY 10023
Tel: (212) 642-5904
Fax: (212) 874-6018

Phillips Bayswater
10 Salem Road
London W2 4BU, England
Tel: 011,44-71-229-9090
Fax: 011,44-71-792-9201

Theriault's
P.O. Box 151
Annapolis, MD 21404
(301) 224-3655

The Toy Soldier Gallery Inc.
24 Main Street
Highland Falls, NY 10928
(914) 446-6731

Weinheimer Auktionshaus
Karlsruher Strasse 218
D 6940 Weinheim B,
Germany
Tel: 011,49-6-2011-5997
Fax: 011,49-6-20118-2891

Elastolin: 10-centimeter English soldiers; value $40 each.

Elastolin: 10-centimeter English soldiers; value $30 each.

Dealers

✤ This is a list of some of the main dealers in old and new old toy soldiers in the United States and England.

Hank Anton
92 Swain Avenue
Meriden, CT 06450
(203) 235-0897
Mail auction: Dime-store
figures

Brunton's Barracks
415 S. Montezuma
Prescott, AZ 86303
(602) 778-1915
Mail order: Old and new old
toy soldiers shop. By
appointment.

Jeanne Burley
34 D. Upper Montagu Street
London W1, England
Mail order: Hollow-cast
British military and civilian

Burlington Antique Toys
1082 Madison Avenue
New York, NY 10028
(212) 861-9708
Shop: Old and new old toy
soldiers

Classic Toys
69 Thompson Street
New York, NY 10012
(212) 941-9129
Shop: Old and new old toy
soldiers

Dutkin's Collectibles
1019 West Route 70
Cherry Hill, NJ 08002
(609) 428-9559
Mail order: Unpainted casts
and molds

Excalibur Hobbies Ltd.
63 Exchange Street
Malden, MA 02148
(617) 322-2959
(617) 322-7910
Shop: Old and new old toy
soldiers

Gettysburg Toy Soldiers
415 Baltimore Street
Gettysburg, PA 17325
(717) 337-3151
Shop: Mainly new old toy
soldiers

Tom Glassic
1030 S. Pine Drive
Bailey, CO 80421
(303) 838-5071
Mail order: Britains

Gordon's Soldiers
44 Page Road
Bow, NH 03304
(603) 224-3924
Mail order: New old toy
soldiers

Tony and Jacki Grecco
P.O. Box 3490
Poughkeepsie, NY 12603
Tel: (914) 462-8829
Fax: (914) 362-1746
Mail order: Mainly dime-store
and composition

The Hobby Chest
1826 Glenview Road
Glenview, IL 60025
(708) 729-5487
Shop: New old toy soldiers

King and Country
Shop 362 The Mall
Pacific Place
Queensway, Hong Kong
Tel: 011,8852-525-8603
Fax: 011,8852-877-9012
Shop: New old toy soldiers

Kings X Toy Soldiers
206 Alamo Plaza
San Antonio, TX 78205
(512) 226-7000
Shop: Mainly new old toy
soldiers, some old

London Toy Soldiers
47 Tottenham Lane
London N8 9BD, England
Tel: 011,44-81-340-2184
Fax: 011,44-81-340-0457
Store: Mainly Britains, new
and old

The March of Time
Bertel Bruun
P.O. Box 400
Westhampton, NY 11977
Tel: (516) 288-6597
Fax: (516) 288-0581
Mail order: Composition,
Heyde and Britains

*Elastolin:
7-centimeter
English soldiers,
values $35,
$65, and $40.*

Tom Mayfield
3232 Cobb Parkway, Suite 170
Atlanta, GA 30339
(404) 352-5130
Mail auction: Composition
soldiers

Memorable Things
31 W. Allegheny Avenue
Towson, MD 21204
(410) 825-3117
Shop: Old toy soldiers

Mickey's House of Soldiers
601 Capella Street
Sunnyvale, CA 94086
(408) 739-1073
Shop: Old and new old toy
soldiers

Midwest Miniatures
4645 Lilac Avenue
Glenview, IL 60025
(708) 296-5465
Mail order: Mainly Mignot
and Britains

K. Warren Mitchell
1008 Forward Pass
Patashala, OH 43062
(614) 927-1661
Mail order: Britains and dime-
store

Stephen Naegel
Stand B23, Lower Ground
Floor
Grays Antique Market
1-7 Davies Mews
London W1Y 1AR, England
Tel: 011,44-71-491-3066
Fax: 011,44-71-493-9344
Shop: Old and new old toy
soldiers

**Northern Toy Soldiers
(G.M. Haley)**
Unit 5, 28 Bridgegate
Hebden Bridge
West Yorks HX7 8EX,
England
011,44-422-842-484

The Old Toy Soldier Home
977 South Santa Fe Avenue
Suite 11
Vista, CA 92083
(619) 758-5481
Shop: Old and new old toy
soldiers

Pageantry Products
8571 Harrison Way
Buena Park, CA 90620
Tel: (714) 995-5867
Fax: (714) 995-4635
Mail order: New old toy
soldiers

The Quarter Master
722 Hurlingham Avenue
San Mateo, CA 94402-1030
Mail order: New old toy
soldiers

The Red Lancer
Dept. TSR, P.O. Box 8056
Mesa, AZ 85214
(602) 964-9662
Mail order: Old toy soldiers,
especially Britains

Roscha Mail Order
22 Leroy Street
New York, NY 10014
(212) 242-4353
Mail order: Mainly Britains
and composition but also
others

Phil Savino
Rt. 2, Box 76
Micanopy, FL 32667
Mail order auction: Dime-
store and Europeans

George Schulz
Bindestr. 24
47228 Duisburg 14, Germany
Mail order: Composition

Second Childhood
283 Bleecker Street
New York, NY 10014
(212) 989-6140
Shop: Old toy soldiers

Shiloh Station
834 E. New Haven Avenue
Melbourne, FL 32901
(407) 951-0515
Shop: Old and new old toy
soldiers, military books

The Ship and Soldier Shop
55 Maryland Avenue
Annapolis, MD 21401
(410) 268-1141
Shop: Old and new old toy
soldiers

Rodger C. Smith
3213 W. Liberty Avenue
Pittsburgh, PA 15216
Tel: (412) 561-1001
Fax: (412) 561-9234
Mail order: New old toy
soldiers

Elastolin: 7-centimeter English soldiers; value $40 each.

Stone Castle Imports
804 N. Third Street
P.O. Box 141
Bardstown, KY 40004
Tel: (502) 897-0207
Fax: (502) 897-6415
Mail order: Imperial
production

TNC
318 Churchill Court
Elizabethtown, KY 42701
(502) 765-5035
Mail order: New old toy
soldiers

Toys and Soldier's Museum
1100 Cherry Street
Vicksburg, MS 39180
(601) 636-0741
Shop and Museum: Mainly
new old toy soldiers in shop

The Toy Soldier Collection
302 So. Main Street
Galena, IL 61036
(815) 777-0383
Shop: Mainly new old toy
soldiers

The Toy Soldier Company
100 Riverside Drive
New York, NY 10024
(201) 433-2370
Mail order: Mainly new old
toy soldiers and plastics, but
some old toy soldiers

The Toy Soldier Exchange
No. 3, The Burlington
Arcade
380 So. Lake Avenue
Pasadena, CA 91101
(818) 356-9780
Shop: Old and new old toy
soldiers

The Toy Soldier Gallery
24 Main Street
Highland Falls, NY 10928
(914) 446-6731
Shop: Old toy soldiers, new
old toy soldiers, military
books

The Toy Soldier Shoppe
The Barn at Stonewood
17700 W. Capitol Drive
Brookfield, WI 53045
(414) 781-6424
Shop: Mainly new old toy
soldiers

Toy Troops
16928 Bolsa Chica Road
Huntington Beach, CA
92649
(714) 846-8486
Shop: New old toy soldiers

Tradition
2 Shepherd Street
Mayfair
London W1Y 7LN, England
Tel: 011,44-71-493-7452
Fax: 011,44-71-355-1224
Shop: Tradition (new old toy
soldiers)

The Trumpeting Angel
76564 W. Bancroft Street
Toledo, OH 43617
(419) 841-4523
Mail order: New old toy
soldiers

Under Two Flags
4 St. Christopher's Place
Vigmore Street
London W1, England
011,44-71-935-6934
Shop: New old toy soldiers

Joe Wallis
P.O. Box 7422
Silver Spring, MD 20910-
7422
Mail order: Britains

Elastolin: 7-centimeter English with Tipp Co. cannon; value: cannon $100; figures $40 each.

Elastolin: 7-centimeter English motorcycle; value $250.

Bibliography

Asquith, S. *The Collector's Guide to New Toy Soldiers.* Argus Books, Hemel Hampstead: Argus Books, 1991. Lists all the New Old Toy Soldier lines up to 1990.

Fontana, Dennis. *War Toys 2. The Story of Lineol.* London: Cavendish Books, 1992. History and description of many aspects of this famous composition maker.

Garratt, J.G. *The World Encyclopedia of Model Soldiers.* Woodstock, NY: Overlook Press, 1981. Excellent although not always correct encyclopedia of the subject.

Greenhill, Peter. *Heraldic Miniature Knights.* Lewes, England: Guild of Master Craftsman Publications, Inc., 1993. Beautifully illustrated book covering Courtenay, Ping, and Des Fontaines figures.

Johnson, P. *Toy Armies.* Garden City, New York: Doubleday & Company, Inc., 1982. Well-written overview based on the Forbes collection.

Joplin, N. *British Toy Figures. 1900 to the present.* Poole: Arms and Armor Press, 1987. Guide covering British non-military figures.

Joplin, N. *The Great Book of Hollow-cast Figures.* London: New Cavendish Books, 1993. Beautiful and exhaustive book on non-Britains hollow casts.

Kearton, G. *The Collectors' Guide to Plastic Toy Soldiers, 1947-1987*. Bolton, England: Ross Anderson Publications, 1987. A guide to plastic soldiers. Useful but not attractive.

Kilian, J. and C. Griffin. *G.I. Joe Collectibles*. Dayton: Tomart Publications, 1993.

Knöbel, R., H. Knöbel, and H. Sieg. *Uniforms of the World*. New York: Charles Scribner's Sons, 1980. Exhaustive handbook on uniforms 1700-1937.

Kurtz, H.I. and B.R. Ehrlich. *The Art of the Toy Soldier*. New York: Abbeville Press, 1987. An overview of the subject with gorgeous photographs.

Leimbweber, V. *Die Kleine Figur. Geschichte in Masse & Zinn*. Statiche Kunstsamlungen Kassel. Kassel: 1985. Description of German flatmakers and German makers of composition figures. In German.

Nevins, Edward. *Forces of the British Empire—1914*. Arlington, VA: Vandemere Press, 1992. Guide to the units, uniforms, and badges of the British Empire in 1914.

O'Brien, R. *The Barclay Catalog Book, 1986*. Obtainable from the author: 135 Stephenburg Rd., RD 2, Port Murray, NJ 07865. Materials from the Barclay Archives.

O'Brien, R. *The Second Catalog Book*. Obtainable from the author: 135 Stephenburg Rd., RD 2, Port Murray, NJ 07865. Toy soldier catalog from Manoil, Warren, All-Nu, Authenticast, Beton, Grey Iron, and Barclay.

O'Brien, R. *Collecting Toy Soldiers No. 2*. Books Americana, 1992. Exhaustive list and price guide, especially comprehensive in regard to U. S. manufacturers.

Opie, F. *British Toy Soldiers. 1893 to the present*. London: Arms and Armor Press, 1985. Guide covering British military figures.

Opie, F. *Britains Toy Soldiers 1893-1932*. New York: Harper & Row, 1985. Beautiful and exhaustive volume on the subject.

Opie, F. *Collecting Toy Soldiers*. London: Collins, 1987. Beautifully produced overview of the subject.

Opie, F. *Toy Soldiers*. *Phillips Collectors' Guides*. London: Boxtree, 1989. A survey of the field with special emphasis on Britains and auctions.

Polaine, R. and D. Hawkins. *War Toys 1. The story of Hausser-Elastolin*. London: Cavendish Books, 1992. History and description of many aspects of this famous composition maker.

Roer, H.H. *Bleisoldaten*. Munich: Callwey, 1981. Well-illustrated book on lead soldiers with emphasis on German manufacturers. In German.

Rose, A. *Toy Soldiers*. London: Salamander Books, Ltd., 1985. Excellent photographic guide to toy soldiers. All color.

Wallis, Joe. *Regiments of All Nations, Britains Ltd. Lead Soldiers, 1946-66*. Joe Wallis. P.O. Box 7422, Silver Spring, MD 20910-7422. 1981. Just as the title says. A full guide to postwar hollow-cast Britains.

Wallis, J. *Armies of the World, Britains Ltd. Lead Soldiers, 1925-1941*. Joe Wallis, P.O. Box 7422, Silver Spring, MD 20910-7422. Just as the title says. A full guide to Britains production in the period covered.

Tipp Co.: Führerwagen; value $2,000. Photograph courtesy of Henry Kurtz, Ltd.

Index

The CONFIDENT COLLECTOR™
KNOWS THE FACTS
Each volume packed with valuable information that
no collector can afford to be without

THE OVERSTREET COMIC BOOK
PRICE GUIDE, 24th Edition
by Robert M. Overstreet   77854-8/$15.00 US/$17.50 Can

THE OVERSTREET COMIC BOOK
GRADING GUIDE, 1st Edition
by Robert M. Overstreet and Gary M. Carter   76910-7/$12.00 US/$15.00 Can

• • •

FINE ART
Indentification and Price Guide, 2nd Edition
by Susan Theran   76924-7/$20.00 US/ $24.00 Can

BOTTLES
Identification and Price Guide, 1st Edition
by Michael Polak   77218-3/$15.00 US/$18.00 Can

ORIGINAL COMIC ART
Identification and Price Guide, 1st Edition
by Jerry Weist   76965-4/$15.00 US/$18.00 Can

COLLECTIBLE MAGAZINES
Identification and Price Guide, 1st Edition
by David K. Henkel   76926-3/$15.00 US/$17.50 Can

COSTUME JEWELRY
Identification and Price Guide, 2nd Edition
by Harrice Simons Miller   77078-4/$15.00 US/$18.00 Can